Metamodeling for Variable Annuities

CHAPMAN & HALL/CRC
Financial Mathematics Series

Aims and scope:
The field of financial mathematics forms an ever-expanding slice of the financial sector. This series aims to capture new developments and summarize what is known over the whole spectrum of this field. It will include a broad range of textbooks, reference works and handbooks that are meant to appeal to both academics and practitioners. The inclusion of numerical code and concrete real-world examples is highly encouraged.

Series Editors

M.A.H. Dempster
Centre for Financial Research
Department of Pure Mathematics and Statistics
University of Cambridge

Dilip B. Madan
Robert H. Smith School of Business
University of Maryland

Rama Cont
Department of Mathematics
Imperial College

Equity-Linked Life Insurance
Partial Hedging Methods
Alexander Melnikov, Amir Nosrati

High-Performance Computing in Finance
Problems, Methods, and Solutions
M. A. H. Dempster, Juho Kanniainen, John Keane, Erik Vynckier

An Introduction to Computational Risk Management of Equity-Linked Insurance
Runhuan Feng

Derivative Pricing
A Problem-Based Primer
Ambrose Lo

Portfolio Rebalancing
Edward E. Qian

Interest Rate Modeling
Theory and Practice, 2nd Edition
Lixin Wu

Metamodeling for Variable Annuities
Guojun Gan and Emiliano A. Valdez

For more information about this series please visit: https://www.crcpress.com/Chapman-and-HallCRC-Financial-Mathematics-Series/book-series/CHFINANCMTH

Metamodeling for Variable Annuities

Guojun Gan
Emiliano A. Valdez

CRC Press
Taylor & Francis Group
Boca Raton London New York

CRC Press is an imprint of the
Taylor & Francis Group, an **informa** business
A CHAPMAN & HALL BOOK

CRC Press
Taylor & Francis Group
6000 Broken Sound Parkway NW, Suite 300
Boca Raton, FL 33487-2742

First issued in paperback 2021

© 2020 by Taylor & Francis Group, LLC
CRC Press is an imprint of Taylor & Francis Group, an Informa business

No claim to original U.S. Government works

ISBN 13: 978-0-367-77955-9 (pbk)
ISBN 13: 978-0-8153-4858-0 (hbk)

Visit the Taylor & Francis Web site at
http://www.taylorandfrancis.com

and the CRC Press Web site at
http://www.crcpress.com

To our students

– Guojun and Emiliano

To my mom Norma and my dad Ponso

– Emiliano

Contents

Preface

Variable annuities are life insurance products that offer various types of financial guarantees. Insurance companies that have a large block of variable annuity business face many challenges. For example, guarantees embedded in variable annuity policies demand sophisticated models for pricing, financial reporting, and risk management. In practice, insurance companies rely heavily on Monte Carlo simulation to calculate the fair market values of the guarantees because the guarantees are complicated and no closed-form valuation formulas are available. One drawback of Monte Carlo simulation is that it is extremely time-consuming and often prohibitive to value a large portfolio of variable annuity contracts because every contract needs to be projected over many scenarios for a long time horizon.

This monograph is devoted to metamodeling approaches, which have been proposed recently in the academic literature to address the computational problems associated with the valuation of large variable annuity portfolios. A typical metamodeling approach involves the following four steps:

1. select a small number of representative variable annuity contracts (i.e., experimental design),

2. use Monte Carlo simulation to calculate the fair market values (or other quantities of interest) of the representative contracts,

3. build a metamodel (i.e., a predictive model) based on the representative contracts and their fair market values, and

4. use the metamodel to estimate the fair market value for every contract in the portfolio.

Using metamodeling approaches can significantly reduce the runtime of valuing a large portfolio of variable annuity contracts for the following reasons: first, building a metamodel only requires using the Monte Carlo simulation model to value a small number of representative contracts; second, the metamodel is usually much simpler and more computationally efficient than Monte Carlo simulation.

This book is primarily written for undergraduate students, who study actuarial science, statistics, risk management, and financial mathematics. It is equally useful for practitioners, who work in insurance companies, consulting firms, and banks. The book is also a source of reference for researchers and graduate students with scholarly interest in computational issues related to variable annuities and other similar insurance products. The methods presented in the book are described in detail and implemented in R, which is a popular language and environment for statistical computing and graphics. Using the R code and datasets that accompany this book, readers can easily replicate the numerical results presented in the book. In addition, readers can modify the R code included in this book for their own use.

This book is divided into three parts. The first part, which consists of Chapters 1, 2, and 3, introduces the computational problems associated with variable annuity valuation, reviews existing approaches, and presents the metamodeling approach in detail. The second part includes Chapters 4, 5, and 6. This part introduces some experimental design methods, which are used to select representative variable annuity contracts. In particular, we describe Latin hypercube sampling, conditional Latin hypercube sampling, and hierarchical k-means for selecting representative policies. The third part includes Chapters 7 to 12 and introduces six metamodels: ordinary kriging, universal kriging, GB2 regression, rank order kriging, linear model with interactions, and tree-based models. These metamodels are predictive models that have been studied in the academic literature for speeding up variable annuity valuation. The dependencies of the chapters are shown in Figure 1. We implement all the experimental design methods and metamodels using a synthetic dataset, which is described in the appendix of this book.

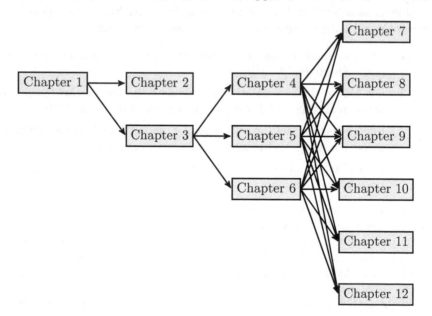

FIGURE 1: Chapter dependencies.

This book has grown from several research projects on variable annuity valuation undertaken by the authors at the University of Connecticut. First, we would like to thank the financial support provided through a CAE (Centers of Actuarial Excellence) grant (i.e., *Applying Data Mining Techniques in Actuarial Science*) from the Society of Actuaries. Second, we would like to acknowledge two CKER (Committee on Knowledge Extension Research) Individual Grants (i.e., *Regression Modeling for the Valuation of Large Variable Annuity Portfolios* and *Valuation of Large Variable Annuity Portfolios with Rank Order Kriging*).

Guojun Gan and Emiliano A. Valdez
Storrs, Connecticut, USA
June 10, 2019

Part I

Preliminaries

Part I

Preliminaries

1

Computational Problems in Variable Annuities

Variable annuities are considered long-term insurance products with investment features that can be used by policyholders to accumulate wealth. These products are attractive investment and retirement vehicles because they contain guarantees such as death benefits and living benefits. In this chapter, we give a brief introduction to variable annuities and the computational challenges associated with their valuation.

1.1 Variable Annuities

A variable annuity (VA) is a contractual agreement between a policyholder and an insurance company. It has been considered a tax deferred retirement product with two phases: the accumulation phase and the payout phase. During the accumulation phase, the policyholder makes either a single lump-sum purchase payment or a series of purchase payments to the insurance company. The policyholder accumulates wealth by investing these payments in the available investment funds provided by the insurance company. During the payout phase, the insurance company makes benefit payments to the policyholder beginning either immediately or at some future specified time. These benefit payments will depend upon the investment performance of the funds selected by the policyholder.

Although VA products were first introduced in the US by TIAA-CREF (Teachers Insurance and Annuities Association - College Retirement Equity Fund) as early as the 1950s, there was a slowing in growth of sales of these products for more than 30 years. See Poterba (1997). State laws restricted insurance companies from issuing insurance products that were backed by assets in their separate accounts. See Maclean (1962).

A major feature of variable annuities is that they come with guarantees. These guarantees can be broadly described in the following two categories:

- The guaranteed minimum death benefit (GMDB). A GMDB provides a guaranteed lump sum to the beneficiary upon the death of the policyholder, regardless of the accumulated value of the policyholder's account.

- The guaranteed minimum living benefit (GMLB). There are several types of GMLB, for example:

  - The guaranteed minimum withdrawal benefit (GMWB). A GMWB guarantees the policyholder the right to withdraw a specified amount during the life of the contract until the initial investment is recovered.
  - The guaranteed minimum accumulation benefit (GMAB). A GMAB provides the policyholder a minimum guaranteed accumulation balance at some future point in time, after a specified waiting period.
  - The guaranteed minimum maturity benefit (GMMB). A GMMB is similar to a GMAB with the exception that the guaranteed accumulation amount is available only at contract maturity.
  - The guaranteed minimum income benefits (GMIB). A GMIB guarantees that the policyholder can convert the greater of the actual account value or the benefit base to an annuity according to a rate predetermined at issue.

Figure 1.1 shows how cash flows circulate between the policyholder and the insurance company. Purchase payments are deposited into the insurance company's separate account; withdrawals are also drawn from the separate account. Investment and other charges are deposited to the insurance company's general account; any payments arising from the product guarantees are drawn out of the general account. In principle, insurers are obligated to set aside reserves for the guarantees that need to be drawn out of the general account. Such is not the case for funds drawn out of the separate account.

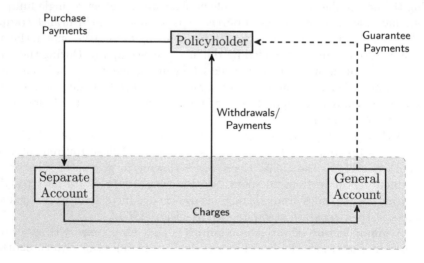

FIGURE 1.1: Cash flows of a typical variable annuity policy.

Table 1.1 illustrates the risks faced by insurance companies with product guarantees from variable annuities with a GMWB. It demonstrates that these companies have to extend these guarantee payments to policyholders especially under poor market conditions. In this example, we have an immediate variable annuity policy with a GMWB rider. The policyholder makes a single lump-sum payment of $100,000 with a maximum annual withdrawal of $8,000. The investment return is negative for many years and with the GMWB, the policyholder is able to continue making annual withdrawal payments until initial investment is recovered. This is despite the policyholder's accumulated funds being exhausted by year 6.

TABLE 1.1: A simple example of the GMWB cash flows, where t, r_t, F_t^-, W_t, F_t^+, B_t, and G_t denote the policy year, the investment return, the account value before withdrawal, the withdrawal amount, the account value after withdrawal, the remaining benefit, and the guarantee payoff, respectively.

t	r_t	F_t^-	W_t	F_t^+	B_t	G_t
1	−10%	90,000	8,000	82,000	92,000	0
2	10%	90,200	8,000	82,200	84,000	0
3	−30%	57,540	8,000	49,540	76,000	0
4	−30%	34,678	8,000	26,678	68,000	0
5	−10%	24,010	8,000	16,010	60,000	0
6	−10%	14,409	8,000	6,409	52,000	0
7	10%	7,050	8,000	0	44,000	**950**
8	r	0	8,000	0	36,000	**8,000**
9	r	0	8,000	0	28,000	**8,000**
10	r	0	8,000	0	20,000	**8,000**
11	r	0	8,000	0	12,000	**8,000**
12	r	0	8,000	0	4,000	**8,000**
13	r	0	4,000	0	0	**4,000**

Its attractive guarantee features, coupled with an uptick in stock prices, helped contribute to a rapid growth in sales of variable annuities in the past two decades. According to the Morningstar Annuity Research Center (The Geneva Association, 2013), the annual VA sales in the United States was $20 billion in 1993. According to LIMRA, VA sales have been averaging over $100 billion a year (see Figure 1.2).

The rapid growth of variable annuities has posed great challenges to insurance companies, especially when it comes to valuing the complex guarantees embedded in these products. The financial risks associated with guarantees embedded in variable annuities cannot be adequately addressed by traditional actuarial approaches (Hardy, 2003). In practice, dynamic hedging is usually adopted by insurers to mitigate the financial risks. Since the guarantees embedded in VA contracts sold by insurance companies are complex, insurers resort to Monte Carlo simulation to calculate the Greeks (i.e., sensitivities of the guarantee values on major market indices) required by dynamic hedging.

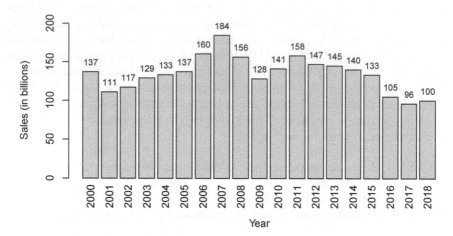

FIGURE 1.2: Sales of variable annuities in the US from 2000 to 2018. Data are obtained from LIMRA Secure Retirement Institute.

However, using Monte Carlo simulation to value a large portfolio of VA contracts is extremely time-consuming (Dardis, 2016): first, long-term projection is usually involved as VA contracts are long-term contracts; second, each contract has its peculiar characteristics and needs to be modeled appropriately.

1.2 Computational Problems Related to Daily Hedging

Dynamic hedging is commonly adopted by insurance companies to mitigate the financial risks associated with variable annuity guarantees. In order to monitor and rebalance the hedge portfolio on a daily basis, insurers need to calculate the Greeks on a set of pre-defined market conditions overnight and interpolate the Greeks in real-time from the pre-calculated Greeks (Gan and Lin, 2017). The Greeks are sensitivities of the fair market values of the guarantees on major market factors. In practice, the bump method (Cathcart et al., 2015) is used to calculate the Greeks. Suppose that there are k market factors. Then the lth partial dollar delta of the guarantees is calculated as follows:

$$
\begin{aligned}
& Delta^{(l)} \\
& = \frac{V_0\left(PA_0^{(1)}, \ldots, PA_0^{(l-1)}, (1+s)PA_0^{(l)}, PA_0^{(l+1)}, \ldots, PA_0^{(k)}\right)}{2s} - \\
& \quad \frac{V_0\left(PA_0^{(1)}, \ldots, PA_0^{(l-1)}, (1-s)PA_0^{(l)}, PA_0^{(l+1)}, \ldots, PA_0^{(k)}\right)}{2s}, \quad (1.1)
\end{aligned}
$$

where s is the shock amount (e.g., 1%) applied to the partial account value and $V_0(\cdot, \cdots, \cdot)$ denotes the fair market value expressed as a function of partial account values. Similarly, the partial dollar rhos are calculated as follows:

$$Rho^{(l)} = \frac{V_0(r_l + s) - V_0(r_l - s)}{2s}, \qquad (1.2)$$

where $V_0(r_l + s)$ is the fair market value calculated based on the yield curve bootstrapped with the lth input rate r_l being shocked up s bps (basis points) and $V_0(r_l - s)$ is similarly defined.

Since the fair market values are calculated by Monte Carlo simulation, calculating the Greeks for a large portfolio of VA policies is extremely time-consuming. For example, suppose that Monte Carlo simulation uses 1,000 risk-neutral scenarios and 360 monthly time steps, the portfolio contains 100,000 VA policies, and 50 pre-defined market conditions are used. Then the total number of cash flow projections is

$$50 \times 1,000 \times 12 \times 30 \times 100,000 = 1.8 \times 10^{12}.$$

Suppose that a computer can process 200,000 cash flow projections per second. Then it would take this computer

$$\frac{1.8 \times 10^{12} \text{ projections}}{200,000 \text{ projections/second}} = 2500 \text{ hours}$$

to process all the cash flows.

Gan and Valdez (2017b) implemented a simple Monte Carlo simulation model to calculate the fair market values of a portfolio of 190,000 synthetic variable annuity policies. As reported in Gan and Valdez (2017b), it would take a single CPU about 108.31 hours to calculate the fair market values for the portfolio at 27 different market conditions.

1.3 Computational Problems Related to Financial Reporting

To reflect the effect of dynamic hedging in quarterly financial reporting, insurers usually employ a stochastic-on-stochastic (i.e., nested simulation) framework. Stochastic-on-stochastic simulation is also used to validate the performance of dynamic hedging. Stochastic-on-stochastic simulation involves two levels of simulation. At the first level, an outer loop is simulated and at the second level, the inner loop is simulated (Reynolds and Man, 2008). Figure 1.3 shows the basic structure of nested simulation. The outer loop involves projecting the VA liabilities along real-world scenarios, which reflect realistic assumptions about the market. At each node of an outer-loop scenario, the

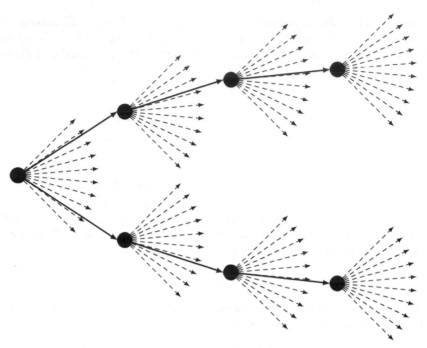

FIGURE 1.3: A sketch of stochastic-on-stochastic simulation. The outer-loop is denoted by solid arrows and the inner-loop is denoted by dashed arrows.

liabilities are projected using a large number of simulated risk-neutral paths, which reflect unrealistic assumptions that investors are risk-neutral.

Stochastic-on-stochastic valuation is extremely computationally demanding (Reynolds and Man, 2008; Gan and Lin, 2015). Consider a stochastic-on-stochastic simulation model with 1,000 outer loops, 1,000 inner loops, and 360 monthly time steps. The total number of cash flow projections for a portfolio of 100,000 VA policies is

$$100,000 \times 1,000 \times 1,000 \times 360 \times 360/2 = 6.48 \times 10^{15}.$$

Suppose that a computer can process 200,000 cash flow projections per second. It would take the computer

$$\frac{6.48 \times 10^{15} \text{ projections}}{200,000 \text{ projections/second}} \approx 1027 \text{ years}$$

to process all the cash flows of the portfolio.

Gan and Valdez (2018b) implemented a simple stochastic-on-stochastic Monte Carlo simulation model to calculate the fair market values of a portfolio of 190,000 synthetic variable annuity policies. As reported in Gan and Valdez (2018b), it would take a single CPU about 3 years to run the stochastic-on-stochastic simulation model on this portfolio.

1.4 Summary

Variable annuities are insurance products with investment features. In this chapter, we gave a brief introduction to variable annuities and their main features. We also discussed two computational challenges associated with the valuation of large variable annuity portfolios. In particular, the first computational challenge is related to the daily hedging of variable annuities that requires calculating the Greeks. The second computational challenge is associated with the use of nested simulation for financial reporting. For more information about variable annuities, readers are referred to Hardy (2003), Ledlie et al. (2008), The Geneva Association (2013), and Feng (2018).

2

Existing Approaches

Existing approaches for addressing the computational problems associated with variable annuity valuation can be divided into two categories: hardware approaches and software approaches. Hardware approaches speed up the valuation process by using hardware. In hardware approaches, multiple central processing units (CPUs) or cores are used to value a portfolio of variable annuity policies in parallel. The policies are divided into small groups, each of which is handled by a CPU/core. Recently, graphics processing units (GPUs), which are massively-parallel processors, have been used to value variable annuities (Phillips, 2012). By contrast, software approaches speed up the valuation by using mathematical models/algorithms. In this chapter, we give an overview of some software approaches that have been proposed to address the computational problems.

2.1 Scenario Reduction

Risk-neutral scenarios, which reflect unrealistic assumptions of investors, are often used in Monte Carlo simulation. When Monte Carlo is used to value a portfolio of variable annuity policies, the payoffs of the guarantees embedded in each policy are calculated at each time step of each risk-neutral scenario. Cutting down the number of scenarios is one way to reduce the runtime of Monte Carlo simulation. Rosner (2011) compared several scenario reduction methods, which are introduced in the following subsections.

2.1.1 Scenario Ranking

Scenario ranking is a technique used by insurance companies to avoid running unnecessary scenarios when calculating tail risk measures such as CTEs (see Remark 2.1). For example, when 1,000 scenarios are used in Monte Carlo simulation, then $CTE_{0.7}$ is calculated based on the worst 300 scenarios among the 1,000 scenarios. However, insurance companies need to run all the 1,000 scenarios to determine the worst 300 scenarios.

Remark 2.1. *The conditional tail expectation (CTE) is a risk measure commonly used by insurance companies to determine the reserve of variable annuities. Let L denote a loss random variable and $\alpha \in (0, 1)$. Then the $100\alpha\%$ CTE risk measure is defined as follows (Hardy, 2003):*

$$\mathrm{CTE}_\alpha = E[L|L > V_\alpha],$$

where V_α is the α-quantile. Using Monte Carlo simulation with N simulations, CTE_α is approximated by

$$\widehat{\mathrm{CTE}}_\alpha = \frac{1}{N(1-\alpha)} \sum_{j=N\alpha+1}^{N} L_{(j)},$$

where $L_{(1)} \leq L_{(2)} \leq \ldots \leq L_{(N)}$ denote the ordered losses from the N simulations. In other words, CTE_α is approximated by the average of the worst $N(1 - \alpha)$ outcomes provided that $N\alpha$ is an integer. When 1,000 simulations are used in Monte Carlo simulation, for example, $\mathrm{CTE}_{0.7}$ is approximated by

$$\widehat{\mathrm{CTE}}_{0.7} = \frac{1}{300} \sum_{j=701}^{1000} L_{(j)}.$$

The main idea of scenario ranking is to determine which scenarios are the worst 300 without having to run all the policies through them. This is achieved by running a small subset of policies through the full 1,000 scenarios to determine the scenario order. A typical process of using the scenario ranking technique to calculate $\mathrm{CTE}_{0.7}$ is as follows:

1. Select a random subset of policies (e.g., 5% of the full portfolio).
2. Run the subset of policies through the full 1,000 scenarios.
3. Order the results (e.g., fair market values) to identify the worst 300 scenarios.
4. Run all policies in the portfolio through the worst 300 scenarios identified above.
5. Calculate the $\mathrm{CTE}_{0.7}$ based on the results.

The scenario order determined based on the subset of policies might be different from that determined based on the full set of policies. To alleviate this problem for calculating the $\mathrm{CTE}_{0.7}$, the scenario ranking technique can be enhanced with the following modifications:

- Use more sophisticated methods to select the subset of policies. For example, data clustering (Gan, 2011) can be used for this purpose.

- Run the full set of policies through the worst 400 scenarios identified based on the subset and then calculate the $\mathrm{CTE}_{0.7}$ based on the worst 300 of the 400 scenarios. By using more than 300 worst scenarios identified by the subset, it is more likely to capture the 300 worst scenarios based on the full set.

2.1.2 Representative Scenarios

Representative scenarios are scenario reduction techniques that select a subset of representative scenarios from the full set of scenarios based on certain characteristics of the scenarios. Representative scenarios are often used to reduce the runtime of cash flow testing models.

Chueh (2002) proposed an interest rate sampling algorithm that can effectively select representative interest rate scenarios and can be implemented straightforwardly in any complex asset/liability model. In particular, the proposed sampling algorithm is able to select "extreme" scenarios as well as "moderate" ones so that the distribution of the model outputs based on the representative scenarios resembles that based on the full run different percentiles.

To select 100 representative interest rate scenarios from a set of N scenarios, for example, the sampling algorithm proposed by Chueh (2002) works as follows:

1. Choose an arbitrary scenario from the N simulated ones and denote it as Pivot #1.

2. Calculate the distances from Pivot #1 to the remaining $N-1$ scenarios.

3. Let Pivot #2 be the scenario with the largest distance to Pivot #1. Randomly decide among ties.

4. Calculate the distances of the $N-2$ non-pivot scenarios to Pivot #1 and Pivot #2.

5. Assign each of the $N-2$ non-pivot scenarios to the closest of Pivot #1 and Pivot #2.

6. Rank the $N-2$ distances in descending order and let Pivot #3 be the scenario producing the top distance.

7. Repeat the above procedure to select the additional 97 pivot scenarios.

8. Assign a probability of N_k/N to Pivot #k, where N_k is the number of scenarios assigned to Pivot #k.

There are several ways to define the distance between two interest rate scenarios. Let $\mathbf{x} = (x_1, x_2, \ldots, x_T)$ and $\mathbf{y} = (y_1, y_2, \ldots, y_T)$ be two interest rate scenarios, where T is the number of periods. Then the distance between $\mathbf{x}$ and $\mathbf{y}$ can be calculated as

$$D(\mathbf{x}, \mathbf{y}) = \left(\sum_{t=1}^{T} (x_t - y_t)^2 V_t \right)^{\frac{1}{2}}, \qquad (2.1)$$

where $V_t \in (0,1)$ is a weight factor that measures the relative importance of the interest rate of each projecting period.

The procedure described above is similar to a clustering algorithm. In fact, data clustering can be used to select representative scenarios (Rosner, 2011). The procedure can also be extended to include equity/bond scenarios (Longley-Cook, 2003).

2.1.3   Importance Sampling

Importance sampling refers to a type of scenario reduction technique that samples scenarios based on their relative importance to the overall result. If some scenarios are more critical to the overall result, then those scenarios are sampled with increased frequencies and reduced weights to produce a final result with no bias.

In importance sampling, the scenarios are first ordered from worst to best based on some specified criterion such as interest rate accumulation factors. Then the bad scenarios are more heavily sampled with reduced weights and good scenarios are more lightly sampled with increased weights. Table 2.1 shows an example of importance sampling given in Rosner (2011). In this example, the ordered scenarios are divided into three groups. The group of worst scenarios is heavily sampled. The group of moderate scenarios is moderately sampled. The group of best scenarios is lightly sampled.

The runtime of importance sampling is similar to that of scenario ranking and representative scenarios. Since importance sampling is a variance reduction technique, it has the potential to increase the accuracy of tail risk measures such as CTEs.

2.1.4   Curve Fitting

Curve fitting is a type of scenario reduction technique that uses a subset of scenarios to approximate the distribution of the model outputs. The approximated distribution can be used to determine the mean or tail metrics of the model outputs.

An example of curve fitting is to use the normal distribution to calculate $\text{CTE}_{0.7}$ for fair market values of a portfolio of variable annuity policies. First, we apply curve fitting to fit a normal distribution to the distribution of the fair market values by determining the mean μ and the standard deviation σ. Second, we calculate the $\text{CTE}_{0.7}$ by the following formula (see Remark 2.2):

$$\text{CTE}_{0.7} = \frac{1}{0.3} \int_{0.7}^{1} V_u \, du, \qquad (2.2)$$

where V_u is the u-quantile of the distribution.

Remark 2.2. *The CTE risk measure is also referred to as the tail-Value-at-Risk or TVaR. Let X be a loss random variable with probability density*

TABLE 2.1: An example of importance sampling.

Scenario Order	Weight
1	4.80%
2	4.80%
3	4.80%
4	
5	14.30%
6	
7	
8	14.30%
9	
10	
11	14.30%
12	
13	
14	
15	
16	
17	42.90%
18	
19	
20	
21	

function $f(x)$. Then CTE_α *or* TVaR_α *can be calculated as (Klugman et al., 2012, p42):*

$$\text{CTE}_\alpha = \text{TVaR}_\alpha = \frac{\int_{V_\alpha}^{\infty} x f(x)\, \mathrm{d}x}{1 - \alpha} = \frac{\int_{\alpha}^{1} V_u\, \mathrm{d}u}{1 - \alpha}, \qquad (2.3)$$

where V_u is the 100u% quantile of the distribution.

Applying curve fitting to a risk measure involves the following steps:

1. Use a subset of scenarios to create model outputs (i.e., fair market values of a portfolio of variable annuity policies).

2. Select an appropriate statistical distribution or a suitable mix of distributions.

3. Choose an appropriate objective function, an optimization algorithm, and constraints for the optimization algorithm.

4. Fit the select distribution to the data by minimizing the objective function.

5. Use the fitted distribution to calculate relevant risk measures.

If the distribution fits the data well, then the risk measures can be calculated accurately without bias.

2.1.5 Random Sampling

Random sampling is a scenario reduction technique used to reduce the runtime of variable annuity valuation. Suppose that a portfolio of n variable annuity policies is valued with $1,000$ risk-neutral scenarios. A full run involves projecting the cash flows of each policy along each of these 1,000 scenarios. Mathematically, the fair market value of the ith policy in the portfolio is calculated as

$$\text{FMV}_i = \frac{1}{1000} \sum_{k=1}^{1000} \text{PV}_k, \tag{2.4}$$

where PV_k is the present value of the cash flows projected along the kth scenario.

In random sampling, a subset of m (m is typically much smaller than 1000) scenarios is randomly selected for each policy. The fair market value is calculated as

$$\widetilde{\text{FMV}}_i = \frac{1}{m} \sum_{k \in C_i} \text{PV}_k, \tag{2.5}$$

where C_i denotes a set of m randomly selected scenarios from the 1,000 scenarios. It is important to note that $C_1, C_2, \ldots, C_n$ are different random subsets of scenarios.

While the fair market values $\widetilde{\text{FMV}}_i$ for individual policies are usually not comparably accurate, the total fair market value of the portfolio is very close, i.e.,

$$\sum_{i=1}^{n} \text{FMV}_i \approx \sum_{i=1}^{n} \widetilde{\text{FMV}}_i.$$

If we are only interested in the total fair market value of the portfolio, then random sampling can be used to reduce the runtime.

2.2 Inforce Compression

In Section 2.1, we introduced scenario reduction techniques to reduce the runtime of actuarial valuation models. In this section, we introduce inforce compression techniques, which can also reduce the runtime by decreasing the number of insurance policies that are valued by the models.

Remark 2.3. *In insurance, the word "inforce" refers to an active block of business. For example, an inforce can be a portfolio of variable annuity policies.*

2.2.1 Cluster Modeling

Cluster modeling is an inforce compression technique that is similar, in some sense, to representative scenarios introduced in the previous section. In cluster modeling, representative policies are selected to resemble the characteristics of the full inforce.

A typical cluster modeling technique involves the following steps:

1. Convert a portfolio of policies into a numerical matrix such that each row represents a policy.

2. Normalize each column of the matrix by using an appropriate normalization method (e.g., z-score normalization, minimax normalization).

3. Apply a clustering algorithm (e.g., the k-means algorithm) to divide the rows of the matrix into clusters.

4. For each cluster, find a representative policy in the cluster that is closest to the cluster center.

5. Scale each representative policy to represent its cluster. For example, the dollar amounts (e.g., account values) of a representative policy are scaled to the total dollar amounts of all policies in its cluster.

6. Use the set of representative policies in place of the portfolio to create model outputs.

Cluster modeling has the potential to reduce the runtime of a valuation model significantly. For example, a portfolio of 500,000 policies can be compressed into a set of 100 representative policies. The runtime can be reduced to a factor of 1/5000 if we ignore the overhead such as time used to read and write files. O'Hagan and Ferrari (2017) compared a variety of clustering techniques for inforce compression.

2.2.2 Replicating Liabilities

Replicating liabilities is an inforce compression technique that uses optimization to produce a small subset of policies that resembles the full inforce closely in certain characteristics.

Replicating liabilities consists of the following two major components:

An objective function To produce a subset of policies that matches the full portfolio, an objective function needs to be determined.

Constraints A set of constraints is also required to make sure that the subset of policies resembles the full portfolio in some characteristics.

For example, suppose that we are interested in the average present value of cash flows across the inforce. We can create a subset of policies that matches the full inforce in account values at different cells or bins. This can be formulated as an optimization problem as follows. Let $\mathbf{x}_i$ denote the ith policy and AV_i denote the account value of the ith policy in the portfolio for $i = 1, 2, \ldots, n$, where n is the number of policies in the portfolio. For $i = 1, 2, \ldots, n$ and $j = 1, 2, \ldots, J$, let B_{ij} be an indicator function that the ith policy is in the jth cell/bin, i.e.,

$$B_{ij} = \begin{cases} 1, & \text{if } \mathbf{x}_i \text{ belongs to the } j\text{th cell,} \\ 0, & \text{if otherwise,} \end{cases}$$

where J is the number of cells. Then the subset of policies can be determined by minimizing the following objective function

$$\sum_{i=1}^{n} I_{\{w_i > 0\}} \tag{2.6}$$

subject to

$$\sum_{i=1}^{n} AV_i \cdot B_{ij} \cdot w_i = \sum_{1}^{n} AV_i \cdot B_{ij}, \quad j = 1, 2, \ldots, J, \tag{2.7a}$$

and

$$w_i \geq 0, \quad i = 1, 2, \ldots, n. \tag{2.7b}$$

That is, we minimize the number of policies in the subset such that the constraints in Equation (2.7) are satisfied.

Commonly used cells/bins include gender and age buckets. For example,

$$B_{i1} = \begin{cases} 1, & \text{if } \mathbf{x}_i \text{ is male,} \\ 0, & \text{if otherwise,} \end{cases} \tag{2.8a}$$

$$B_{i2} = \begin{cases} 1, & \text{if the age of } \mathbf{x}_i \text{ is in } (0, 25) , \\ 0, & \text{if otherwise,} \end{cases} \tag{2.8b}$$

and

$$B_{i3} = \begin{cases} 1, & \text{if the age of } \mathbf{x}_i \text{ is in } [25, \infty) , \\ 0, & \text{if otherwise.} \end{cases} \tag{2.8c}$$

The constraints given in Equation (2.7) have solutions. For example, $w_1 = w_2 = \cdots = w_n = 1$ is a solution. However, finding a solution that minimizes Equation (2.6) is not straightforward. Another objective function is

$$\sum_{i=1}^{n} w_i. \tag{2.9}$$

Minimizing Equation (2.9) subject to constraints given in Equation (2.7) is a linear programming problem and can be solved efficiently. In fact, a linear programming algorithm can find a subset of policies such that their weights w_i are positive and the remaining policies' weights are reduced to zeros.

Once a subset of policies with positive weights are identified. These policies are scaled with their weights. The scaled policies are used in place of the full inforce in the valuation model.

2.2.3 Replicated Stratified Sampling

Replicated stratified sampling (RSS) is an inforce compression technique proposed by Vadiveloo (2012) and Sarukkali (2013). RSS extends traditional stratified sampling by using replicated samples in order to eliminate sampling errors. RSS can be used to estimate tail risk metrics at baseline assumptions as well as modified assumptions. In particular, RSS is suitable to estimate the relative change in a risk metric when a shock is applied to the input variables.

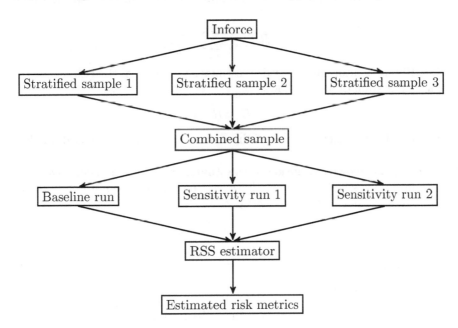

FIGURE 2.1: A sketch of the RSS method with three stratified samples and two sensitivity runs.

Figure 2.1 shows a sketch of the RSS method. Mathematically, RSS works as follows. Let $\mathbf{x}_1, \mathbf{x}_2, \ldots, \mathbf{x}_n$ be the n policies in the inforce. Let k be the sample size and let M be the number of risk classes. For $m = 1, 2, \ldots, M$, the inforce is divided into s strata according to risk class m, where s is the number of strata within the inforce. Each policy is assigned to a stratum. Let

n_j be the number of policies assigned to stratum j. For $m = 1, 2, \ldots, M$, we draw

$$\frac{n_j k}{n}$$

policies from the stratum j created according to risk class m. We denote these policies as $\mathbf{x}_1^{(m)}, \ldots, \mathbf{x}_k^{(m)}$.

Then we let

$$A_m = g(\mathbf{x}_1^{(m)}, \ldots, \mathbf{x}_k^{(m)}, p_1, p_2, \ldots, p_r) \tag{2.10}$$

denote the baseline value of the risk metric, where $p_1, \ldots, p_r$ represent model assumptions (e.g., mortality rates, interest rates, equity returns) and the function g represents the relationship between the inputs and the risk metric. The value of the risk metric under shocked assumptions can be defined similarly as follows:

$$B_m = g(\mathbf{x}_1^{(m)}, \ldots, \mathbf{x}_k^{(m)}, \tilde{p}_1, \tilde{p}_2, \ldots, \tilde{p}_r), \tag{2.11}$$

where $\tilde{p}_1, \ldots, \tilde{p}_r$ denote shocked assumptions. The RSS estimator is defined as

$$\hat{R} = \frac{1}{M} \sum_{m=1}^{M} \frac{B_m}{A_m}. \tag{2.12}$$

The RSS estimator can be used to estimate the value of the risk metric under shocked assumptions as follows:

$$B = \hat{R} \cdot A, \tag{2.13}$$

where A is the baseline value of the risk metric calculated based on the full inforce.

From the above description, we see that the RSS method can save runtime for additional sensitivity runs. However, the full inforce is used to calculate the baseline value of the risk metric.

2.3   Summary

In this chapter, we introduced some existing techniques that can be used to reduce the runtime of stochastic valuation models for variable annuities. In general, the runtime of a stochastic valuation model can be reduced in three ways: reduce the number of scenarios, reduce the number of policies, and reduce the number of time steps in cash flow projections. The techniques introduced in this chapter fall into the first two categories. Reducing the time steps has the potential to reduce runtime of a valuation model but is rarely studied in the literature. For empirical analysis of some of the techniques introduced in this chapter, readers are referred to Rosner (2011). For various algorithms for cluster modeling, readers are referred to O'Hagan and Ferrari (2017).

3

Metamodeling Approaches

In the previous chapter, we introduced some techniques that can be used to address the computational issues associated with variable annuity valuation. Since 2013, metamodeling approaches that originate from system engineering have been proposed and developed in the literature to address the computational problems. In this chapter, we give a broad overview of the metamodeling approach with its major components.

3.1 A General Framework

The term metamodel comes from simulation metamodeling (Friedman, 1996; Barton, 2015). In simulation metamodeling, a metamodel refers to a model of a simulation model, which is usually very complex and computationally intensive. A metamodel is constructed by running a small number of expensive simulations and used in place of the simulation model for further analysis. Since a metamodel is simple and computationally efficient, a metamodeling approach can reduce the runtime of the simulation model significantly.

Remark 3.1. *The concept of metamodels for simulation models was first introduced by Kleijnen (1975). A metamodel is a model of another model. For example, a metamodel can be thought of as a model of an actuarial valuation model (e.g., Monte Carlo simulation).*

Figure 3.1 shows a general framework of metamodeling approaches. In this figure, metamodeling approaches are used to speed up the calculation of the fair market values of guarantees for a large portfolio of variable annuity policies. However, we can use the metamodeling approaches to estimate any quantities of interest (e.g., Greeks or sensitivities of fair market values). From Figure 3.1, we see that a metamodeling approach consists of four major steps:

1. Use an experimental design method to select a small set of representative policies.

21

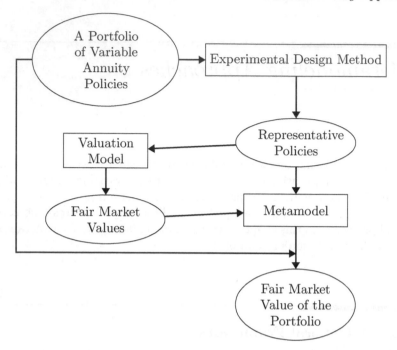

FIGURE 3.1: A general framework of metamodeling approaches. Here we focus on estimating the fair market values of guarantees embedded in variable annuities.

2. Run the valuation model (e.g., Monte Carlo simulation) to calculate the fair market values (or other quantities of interest) of the guarantees for the representative policies.

3. Build a metamodel based on the representative policies and their fair market values.

4. Use the metamodel to predict the fair market values of the guarantees for all policies in the portfolio.

Using metamodeling approaches has the potential to reduce significantly the runtime of valuing a large portfolio of variable annuity policies for the following reasons:

- Building a metamodel only requires using the valuation model to value a small number of representative variable annuity policies.

- The metamodel is usually much simpler and faster than the valuation model.

Metamodeling approaches are similar to inforce compression techniques introduced in Section 2.2 in that both reduce runtime through cutting down the number of policies that are valued by the valuation model. However, metamodeling approaches are more sophisticated than inforce compression techniques because the former adopts metamodels to improve the accuracy of the estimates.

As we can see from Figure 3.1 and the major steps of metamodeling approaches, a metamodeling approach relies on the following two important and interconnected components:

An experimental design method The experimental design method is used to select representative variable annuity policies. Examples of experimental design methods include data clustering algorithms.

A predictive model The predictive model is the metamodel. A wide range of predictive models can be used.

3.2 Literature Review

In this section, we review some metamodeling approaches that have been proposed and studied in the literature for the valuation of variable annuities. In particular, we focus on the two important components of a metamodeling approach: the experimental design method and the predictive model.

TABLE 3.1: Some publications on metamodeling approaches. In the table header, EDM refers to experimental design method.

Publication	EDM	Metamodel
Gan (2013)	Clustering	Kriging
Gan and Lin (2015)	Clustering	Kriging
Gan (2015)	LHS	Kriging
Hejazi and Jackson (2016)	Uniform sampling	Neural network
Gan and Valdez (2016)	Clustering, LHS	GB2 regression
Gan and Valdez (2017a)	Clustering	Gamma regression
Gan and Lin (2017)	LHS	Kriging
Hejazi et al. (2017)	Uniform sampling	Kriging, IDW, RBF
Gan and Huang (2017)	Clustering	Kriging
Xu et al. (2018)	Random sampling	Neural network
Gan and Valdez (2018c)	Clustering	GB2 regression
Gan (2018)	Random sampling	Linear model
Gan et al. (2018)	Clustering	Regression trees
Doyle and Groendyke (2018)	Not disclosed	Neural network

Table 3.1 shows a list of academic papers about using metamodeling ap-

proaches to speed up variable annuity valuation that were published when the book was written. When you are reading this book, you may see other papers that are not shown in this table. All the papers listed in the table differ in the experimental design methods and/or metamodels.

Gan (2013) is perhaps one of the first papers published in academic journals that investigate the use of metamodeling for variable annuity valuation. Gan (2013) applied the k-prototypes algorithm, which is a clustering algorithm proposed by Huang (1998) for mixed-type data, to select representative variable annuity policies and used the ordinary kriging as the metamodel. The dataset used in Gan (2013) is very simple in that only a few types of guarantees are considered.

Gan and Lin (2015) studied the use of metamodeling for the variable annuity valuation under the stochastic-on-stochastic simulation framework. The authors also used the k-prototypes algorithm to select representative variable annuity policies. However, the authors used the universal kriging for functional data (UKFD) as the metamodel because the model outputs under the stochastic-on-stochastic simulation model are functional data.

Gan (2015) proposed to use the Latin hypercube sampling (LHS) method to select representative variable annuity policies. The motivation behind the work of Gan (2015) is that the k-prototypes algorithm is not efficient (i.e., slow) for selecting a moderate number (e.g., 200) of representative variable annuity policies.

Hejazi and Jackson (2016) proposed to use neural networks to speed up the valuation of variable annuity policies. In this paper, the authors ignored the impact of experimental design methods and used uniform sampling to select representative variable annuity policies. In addition, the authors considered metamodeling under a spatial interpolation framework and used neural networks to find an effective distance function.

Gan and Valdez (2016) empirically compared several methods for selecting representative variable annuity policies. In particular, the authors compared random sampling, low-discrepancy sequences, data clustering, Latin hypercube sampling, and conditional Latin hypercube sampling. They found that the clustering method and the Latin hypercube sampling method are comparable in terms of prediction accuracy and are better than other methods such as random sampling.

Gan and Valdez (2017a) investigated the use of copulas to model the dependency of partial dollar deltas. In particular, the authors considered several copulas to model the dependency of partial dollar deltas. The main finding is that the use of copulas does not improve the prediction accuracy of the metamodel because the dependency is well captured by the covariates (i.e., characteristics of variable annuity policies).

Gan and Lin (2017) studied the use of metamodeling to calculate dollar deltas quickly for daily hedging purpose. For that purpose, the authors proposed a two-level metamodeling approach, which involves two metamodels. The experimental design methods for the two metamodels are different:

one metamodel uses Latin hypercube sampling and the other metamodel uses conditional Latin hypercube sampling.

Hejazi et al. (2017) treated the valuation of large variable annuity portfolios as a spatial interpolation problem and investigated several interpolation methods, including the inverse distance weighting (IDW) method and the radial basis function (RBF) method.

Gan and Huang (2017) proposed a data mining framework for the valuation of large variable annuity portfolios. Gan and Huang (2017) used the truncated fuzzy c-means (TFMC) algorithm, which is a scalable clustering algorithm developed by Gan et al. (2016), to select representative variable annuity policies and used the ordinary kriging as the metamodel.

Xu et al. (2018) proposed neural networks as well as tree-based models with moment matching to value large variable annuity portfolios. Doyle and Groendyke (2018) also proposed neural networks for the valuation of large variable annuity portfolios.

Gan and Valdez (2018c) proposed to use the GB2 (generalized beta of the second kind) distribution to model the fair market values. The motivation behind the work of Gan and Valdez (2018c) is that the distribution of the fair market values is highly skewed. Since the GB2 distribution has three shape parameters, it is flexible to fit highly skewed data.

Gan (2018) explored the use of interaction terms in metamodels to improve the prediction accuracy. In particular, the author investigated the use of linear regression models with interaction effects for the valuation of large variable annuity portfolios and found that linear regression models with interactions are able to produce accurate predictions. In this study, the author ignored the impact of experimental design methods and used random sampling to select representative variable annuity policies.

3.3 Summary

In this chapter, we gave a brief introduction to metamodeling approaches for speeding up variable annuity valuation. Metamodeling stems from computer simulation modeling and is not new. Since Kleijnen (1975) first introduced the concept of metamodels for simulation models, many papers on metamodeling and its applications have been published. For example, Kleijnen (2009) presented a review of the kriging metamodel. Friedman (1996) is a book on simulation metamodeling and introduces usage, applications, and methodology of metamodels. Box and Draper (2007) and Das (2014) are two books that cover many topics on response surface methodology, which is related to metamodeling.

Part II

Experimental Design Methods

4

Latin Hypercube Sampling

Latin hypercube sampling (LHS) is a statistical method introduced by McKay et al. (1979) for generating design points from a multidimensional space. These design points are structurally generated to form a Latin hypercube. It has been used to construct simulation experiments and estimate high-dimensional integrals. In this chapter, we introduce the Latin hypercube sampling method for selecting representative variable annuity policies.

4.1 Description of the Method

In statistical sampling, a Latin square is a square grid of sample positions from a two-dimensional space such that there is only one sample in each row and each column. Figure 4.1 shows two examples of Latin squares, each of which contains four sample positions. Each sampling position is marked by "X." From the figure, we see that only one sample is in each row and each column.

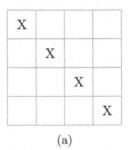
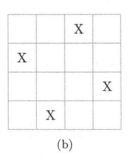

(a)   (b)

FIGURE 4.1: Two examples of Latin squares.

From a statistical sampling point of view, the Latin square shown in Figure 4.1(b) is better than the one shown in Figure 4.1(a) because sample positions in the latter fill the space better than those in the former. The quality of a Latin square can be measured by the minimum distance between its sample positions. The higher the minimum distance, the better the Latin square.

A Latin hypercube is the generalization of a Latin square to a high dimensional space such that each axis-aligned hyperplane contains only a single sample. The quality of a Latin hypercube can also be measured by the minimum distances between its sample positions. When the number of divisions and the number of dimensions increase, the number of Latin hypercubes increases exponentially (McKay and Wanless, 2008). In a three-dimensional space with four divisions in each dimension, for example, the number of Latin hypercubes is

$$64 \times 4! \times (3!)^3 = 331,776.$$

As a result, it is clearly not computationally feasible to find the best Latin hypercube that has the largest minimum distance by enumerating all the Latin hypercubes. One way to find a good Latin hypercube is to generate Latin hypercubes randomly and select the best one from the pool.

The R package `lhs` provides a function called `maximinLHS` that can be used to produce a Latin hypercube sample. This function uses the maximin criterion to select an optimal sample, that is, it optimizes the sample by maximizing the minimum distance between design points. Gan (2015) proposed an LHS method to select representative variable annuity policies by modifying the maximin criterion so that it can handle categorical variables. In what follows, we describe the modified LHS method proposed by Gan (2015).

Let $X = \{\mathbf{x}_1, \mathbf{x}_2, \ldots, \mathbf{x}_n\}$ denote a portfolio of n variable annuity policies. Let each variable annuity policy be characterized by d attributes (e.g., gender, age, account value, etc.). For the sake of simplicity, suppose that the first d_1 attributes are numerical and the remaining $d_2 = d - d_1$ attributes are categorical. For $j = 1, 2, \ldots, d_1$, let L_j and H_j denote the minimum and maximum values of the jth variable, respectively, i.e.,

$$L_j = \min\{x_j : \mathbf{x} \in X\}, \quad H_j = \max\{x_j : \mathbf{x} \in X\}, \tag{4.1}$$

where x_j denotes the jth component of $\mathbf{x}$. For $j = d_1 + 1, d_1 + 2, \ldots, d$, let N_j denote the number of levels (i.e., distinct values) of the jth variable, i.e.,

$$N_j = |\{x_j : \mathbf{x} \in X\}|, \tag{4.2}$$

where $|\cdot|$ denotes the number of elements in a set.

Let $k \geq 2$ be the desired number of design points in a Latin hypercube sample. We first divide the range of each of the d_1 numerical variables into k divisions as follows:

$$I_l = \left(L_j + \left(l - \frac{3}{2}\right)\frac{H_j - L_j}{k-1}, L_j + \left(l - \frac{1}{2}\right)\frac{H_j - L_j}{k-1}\right] \tag{4.3}$$

for $l = 1, 2, \ldots, k$. Then we have

$$\bigcup_{l=1}^{k} I_l = \left(L_j - \frac{H_j - L_j}{2(k-1)}, H_j + \frac{H_j - L_j}{2(k-1)}\right] \subset [L_j, H_j],$$

which indicates that the union of the k divisions covers the whole range of the jth variable. For each of the remaining categorical variables, each category is considered as a division.

Let $\mathcal{H}$ be the set of all grid points from the aforementioned divisions, i.e.,

$$\mathcal{H} = \{(a_1, a_2, \ldots, a_d)\}, \tag{4.4}$$

where for $j = 1, 2, \ldots, d_1$,

$$a_j \in \left\{ L_j + (l-1) \frac{H_j - L_j}{k-1}, l = 1, 2, \ldots, k \right\},$$

and for $j = d_1 + 1, d_1 + 2, \ldots, d$,

$$a_j \in \{A_{jl}, l = 1, 2, \ldots, N_j\},$$

where A_{j1}, A_{j2}, ..., A_{jN_j} are the levels of the jth variable. The number of points contained in $\mathcal{H}$ is

$$|\mathcal{H}| = k^{d_1} \prod_{j=d_1+1}^{d} N_j,$$

which can be a huge number when k is large.

To select an optimal sample with k design points, we proceed as follows. First, we need to define a criterion to measure the quality of a sample. Let H be a subset of $\mathcal{H}$ with k elements. We define the score of H to be the minimum distance between any pairs of distinct points in H, i.e.,

$$S(H) = \min\{D(\mathbf{a}, \mathbf{b}) : \mathbf{a} \in H, \mathbf{b} \in H, \mathbf{a} \neq \mathbf{b}\}, \tag{4.5}$$

where

$$D(\mathbf{a}, \mathbf{b}) = \sum_{j=1}^{d_1} \frac{|a_j - b_j|}{H_j - L_j} + \sum_{j=d_1+1}^{d} \delta(a_j, b_j). \tag{4.6}$$

Here a_j and b_j are the jth components of $\mathbf{a}$ and $\mathbf{b}$, respectively, and $\delta(\cdot, \cdot)$ is defined as

$$\delta(a_j, b_j) = \begin{cases} 0, & \text{if } a_j = b_j, \\ 1, & \text{if } a_j \neq b_j. \end{cases} \tag{4.7}$$

Since we use the maximin criterion, Latin hypercube samples with higher scores are better. An optimal Latin hypercube sample with k design points is defined as

$$H^* = \operatorname*{argmax}_{H \subset \mathcal{H}, |H|=k} S(H). \tag{4.8}$$

Finding an optimal Latin hypercube sample with k design points from $\mathcal{H}$ is not easy as the set $\mathcal{H}$ usually contains a huge number of points. To do that, we randomly generate a set of Latin hypercube samples and select the one with the highest score. We proceed as follows to generate a random Latin hypercube sample:

1. For each $j = 1, 2, \ldots, d_1$, we randomly generate k uniform real numbers from the interval $[0, 1]$ that are denoted by $r_{j1}, r_{j2}, \ldots, r_{jk}$. These numbers are mutually distinct in general. Let $(i_1, i_2, \ldots, i_k)$ be a permutation of $(1, 2, \ldots, k)$ such that

    $$r_{ji_1} < r_{jr_2} < \cdots < r_{jr_k}.$$

    Then we define the first d_1 coordinates of the k design points as

    $$a_{jl} = L_j + (i_l - 1)\frac{H_j - L_j}{k - 1}, \quad j = 1, \ldots, d_1, \, l = 1, \ldots, k.$$

    For each $j = 1, 2, \ldots, d_1$, the coordinates of the k design points in the jth dimension are mutually distinct.

2. For each $j = d_1 + 1, d_1 + 2, \ldots, d$, we randomly generate k uniform integers from $\{1, 2, \ldots, N_j\}$ that are denoted by $i_1, i_2, \ldots, i_k$. Then we define the remaining d_2 coordinates of the k design points as

    $$a_{jl} = A_{ji_l}, \quad j = d_1 + 1, \ldots, d, \, l = 1, \ldots, k,$$

    where $A_{j1}, A_{j2}, \ldots, A_{jN_j}$ are the distinct categories of the jth variable.

We repeat the above procedure to generate many samples and calculate their scores. We select the one with the highest score to be an optimal Latin hypercube sample.

Let $H^* = \{\mathbf{a}_1^*, \mathbf{a}_2^*, \ldots, \mathbf{a}_k^*\}$ be an optimal Latin hypercube sample. The second step of the LHS method is to find k representative variable annuity policies that are close to the k design points in H^*. The policy that is close to $\mathbf{a}_i^*$ is determined by

$$\mathbf{z}_i = \operatorname*{argmin}_{\mathbf{x} \in X} D(\mathbf{a}_i^*, \mathbf{x}), \quad i = 1, 2, \ldots, k,$$

where $D(\cdot, \cdot)$ is defined in Equation (4.6). It is possible that the policies that are closest to two design points are the same.

Another issue with the use of Latin hypercube sampling is how many design points we should use. That is, what is an appropriate value of k? This question is related to the metamodel we will use in the next step. As suggested in Loeppky et al. (2009), the number of design points should be 10 times the number of dimensions of the data.

4.2   Implementation

In this section, we implement the modified LHS method in R. In particular, we implement the following functions:

- A function to generate Latin hypercube samples from an input space.

- A function to calculate the distance matrix of the design points.

- A function to calculate the score of a Latin hypercube sample.

The first function is implemented in R as follows:

```
 1 mylhs <- function(k, L, H, A) {
 2   # continuous part
 3   d1 <- length(L)
 4   mR <- matrix(runif(k*d1), nrow=k)
 5   mOrder <- apply(mR, 2, order)
 6   mN <- matrix(L, nrow=k, ncol=d1, byrow=T) + (mOrder -
          1) * matrix(H-L, nrow=k, ncol=d1, byrow=T) / (k-1)
 7
 8   # categorical part
 9   d2 <- length(A)
10   mC <- matrix(0, nrow=k, ncol=d2)
11   for(j in 1:d2) {
12     mC[,j] <- sample(1:A[j], k, replace=T)
13   }
14   return(cbind(mN,mC))
15 }
```

The function is named mylhs to differentiate it from other similar functions. In this function, we first generate the continuous part by creating permutations of random real numbers. Then we generate the categorical part by sampling from the categories. The function has four arguments: k, L, H, and A. The first argument specifies the number of design points. The arguments L and H specify the minimum and maximum values of the continuous variables. The last argument A is a vector of number of categories.

We can test the function mylhs as follows:

```
 1 > L <- c(1,10)
 2 > H <- c(10,20)
 3 > A <- c(2, 5)
 4 > set.seed(1)
 5 > m1 <- mylhs(20,L,H,A)
 6 > table(m1[,3])
 7
 8   1   2
 9   9  11
10 > table(m1[,4])
11
12   1  2  3  4  5
13   1  7  3  3  6
14 >
15 > set.seed(2)
16 > m2 <- mylhs(20,L,H,A)
```

```
17 > table(m2[,3])
18
19  1   2
20  7  13
21 > table(m2[,4])
22
23 1 2 3 4 5
24 4 6 4 3 3
```

In the above code, we created two samples, each of which contains 20 design points. Each design point has two numerical components and two categorical components. The frequencies of the categorical components are also shown. The continuous components of the two Latin hypercube samples are shown in Figure 4.2. From the figure, we see that the continuous part of the first sample is better than that of the second one as the design points of the first sample fills the space better.

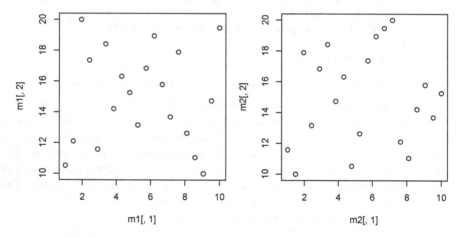

FIGURE 4.2: Continuous part of two Latin hypercube samples.

The second function is implemented as follows:

```
1 calDist <- function(mS, L, H) {
2   k <- nrow(mS)
3   d1 <- length(L)
4   d2 <- ncol(mS) - d1
5
6   mD <- matrix(0, nrow=k, ncol=k)
7   mN <- mS[,1:d1]      # continuous part
8   mC <- mS[,-(1:d1)]   # categorical part
9   for(j in 1:k) {
10     mD[,j] <- abs(mN - matrix(mN[j,], nrow=k, ncol=d1,
           byrow=T)) %*%
11         matrix( 1/(H-L), ncol=1) +
```

```
12            apply(sign(abs( mC - matrix(mC[j,], nrow=k, ncol=
              d2, byrow=T) )), 1, sum)
13     }
14
15     return(mD)
16 }
```

The function is named `calDist` and has three arguments. The first argument is a matrix, which represents a Latin hypercube sample created by the function `mylhs`. The second and the third arguments are the vectors of lower and upper bounds of the continuous variables. For efficiency purposes, we used vectorization to calculate the distances.

To test the function `calDist`, we use it to calculate the distance matrices of the two Latin hypercube samples we generated above as follows:

```
 1 > mD1 <- calDist(m1, L, H)
 2 > mD2 <- calDist(m2, L, H)
 3 > mD1[1:5,1:5]
 4         [,1]      [,2]      [,3]      [,4]      [,5]
 5 [1,] 0.000000 1.6842105 1.421053 1.4210526 1.736842
 6 [2,] 1.684211 0.0000000 2.105263 0.2631579 2.421053
 7 [3,] 1.421053 2.1052632 0.000000 1.8421053 1.315789
 8 [4,] 1.421053 0.2631579 1.842105 0.0000000 2.157895
 9 [5,] 1.736842 2.4210526 1.315789 2.1578947 0.000000
10 > mD2[1:5,1:5]
11         [,1]      [,2]      [,3]      [,4]      [,5]
12 [1,] 0.000000 2.736842 2.052632 2.789474 2.368421
13 [2,] 2.736842 0.000000 1.315789 1.894737 1.631579
14 [3,] 2.052632 1.315789 0.000000 2.210526 2.315789
15 [4,] 2.789474 1.894737 2.210526 0.000000 3.526316
16 [5,] 2.368421 1.631579 2.315789 3.526316 0.000000
```

The distance matrices are 20×20 matrices. In the above output, we show only a portion of the matrices. From the output, we see that the distance matrices are symmetric. The diagonal elements are zeros because the distance between a design point and itself is zero.

The third function is implemented as follows:

```
1 calScore <- function(mD) {
2   ind <- lower.tri(mD)
3   return(min(mD[ind]))
4 }
```

This implementation is straightforward. We first use the R function `lower.tri` to get the indices of the lower triangle of the distance matrix. Then we find the minimum distance from the lower triangle of the distance matrix.

To test the score function, we use it to calculate the score of the two distance matrices:

```
1 > calScore(mD1)
2 [1] 0.2631579
3 > calScore(mD2)
4 [1] 0.2105263
```

The output shows that the first Latin hypercube sample has a higher score and is preferable to the second one.

4.3   Examples

In the previous section, we implemented functions to generate Latin hypercube samples and calculate their scores. Now we apply these functions to the synthetic inforce described in Section A.1.

First, we need to load the inforce into R. Suppose that the current working directory contains the dataset. Then we load the dataset as follows:

```
 1 > inforce <- read.csv("inforce.csv")
 2 > summary(inforce[,1:10])
 3     recordID            survivorShip gender
 4   Min.    :      1   Min.    :1    F: 76007
 5   1st Qu.:  47501   1st Qu.:1    M:113993
 6   Median :  95001   Median :1
 7   Mean    :  95001   Mean    :1
 8   3rd Qu.:142500   3rd Qu.:1
 9   Max.    :190000   Max.    :1
10
11     productType         issueDate           matDate
12   ABRP    :  10000   Min.    :36526   Min.    :42005
13   ABRU    :  10000   1st Qu.:37773   1st Qu.:45566
14   ABSU    :  10000   Median :39052   Median :47088
15   DBAB    :  10000   Mean    :39064   Mean    :47097
16   DBIB    :  10000   3rd Qu.:40330   3rd Qu.:48639
17   DBMB    :  10000   Max.    :41609   Max.    :52201
18   (Other):130000
19     birthDate           currentDate         baseFee
20   Min.    :18264   Min.    :41791   Min.    :0.02
21   1st Qu.:21002   1st Qu.:41791   1st Qu.:0.02
22   Median :23743   Median :41791   Median :0.02
23   Mean    :23728   Mean    :41791   Mean    :0.02
24   3rd Qu.:26451   3rd Qu.:41791   3rd Qu.:0.02
25   Max.    :29190   Max.    :41791   Max.    :0.02
26
27     riderFee
28   Min.    :0.002500
```

```
29  1st Qu.:0.005000
30  Median  :0.006000
31  Mean    :0.006184
32  3rd Qu.:0.007500
33  Max.    :0.009000
```

In the above output, we see the summary statistics of the first ten variables. Many variables have the same value. For example, the variable baseFee was designed to have identical values.

To select representative policies from the inforce, we only need to consider variables that have different values. We select these variables as follows:

```
 1  > vNames <- c("gbAmt", "gmwbBalance", "withdrawal",
        paste("FundValue", 1:10, sep=""))
 2  >
 3  > age <- with(inforce, (currentDate-birthDate)/365)
 4  > ttm <- with(inforce, (matDate - currentDate)/365)
 5  >
 6  > datN <- cbind(inforce[,vNames], data.frame(age=age,
        ttm=ttm))
 7  > datC <- inforce[,c("gender", "productType")]
 8  > dat <- as.matrix(cbind(datN, data.frame(lapply(datC,as
        .numeric))))
 9  >
10  > summary(datN)
11        gbAmt          gmwbBalance          withdrawal
12  Min.   :  50002   Min.   :       0   Min.   :       0
13  1st Qu.:179759    1st Qu.:       0   1st Qu.:       0
14  Median :303525    Median :       0   Median :       0
15  Mean   :313507    Mean   :   36141   Mean   :   21928
16  3rd Qu.:427544    3rd Qu.:       0   3rd Qu.:       0
17  Max.   :989205    Max.   :  499709   Max.   :  499586
18     FundValue1         FundValue2          FundValue3
19  Min.   :       0   Min.   :       0   Min.   :       0
20  1st Qu.:       0   1st Qu.:       0   1st Qu.:       0
21  Median :   8299   Median :   8394    Median :   4942
22  Mean   :  26611   Mean   :  26045    Mean   :  17391
23  3rd Qu.:  39209   3rd Qu.:  38463    3rd Qu.:  24251
24  Max.   : 921549   Max.   : 844323    Max.   : 580753
25     FundValue4         FundValue5          FundValue6
26  Min.   :       0   Min.   :       0   Min.   :       0
27  1st Qu.:       0   1st Qu.:       0   1st Qu.:       0
28  Median :   4225   Median :   7248    Median :   8556
29  Mean   :  14507   Mean   :  21041    Mean   :  26570
30  3rd Qu.:  20756   3rd Qu.:  32112    3rd Qu.:  39241
31  Max.   : 483937   Max.   : 494382    Max.   : 872707
32     FundValue7         FundValue8          FundValue9
33  Min.   :       0   Min.   :       0   Min.   :       0
34  1st Qu.:       0   1st Qu.:       0   1st Qu.:       0
```

```
35  Median  :   6602    Median  :   6255    Median  :   5943
36  Mean    :  21506    Mean    :  19990    Mean    :  19647
37  3rd Qu.:  31088    3rd Qu.:  29404    3rd Qu.:  28100
38  Max.    :634819    Max.    :562485    Max.    :663196
39   FundValue10             age                  ttm
40  Min.    :      0    Min.    :34.52    Min.    : 0.5863
41  1st Qu.:      0    1st Qu.:42.03    1st Qu.:10.3425
42  Median  :   6738    Median  :49.45    Median  :14.5123
43  Mean    :  21003    Mean    :49.49    Mean    :14.5362
44  3rd Qu.:  31256    3rd Qu.:56.96    3rd Qu.:18.7616
45  Max.    :599675    Max.    :64.46    Max.    :28.5205
46 > summary(datC)
47  gender          productType
48  F: 76007    ABRP    : 10000
49  M:113993    ABRU    : 10000
50              ABSU    : 10000
51              DBAB    : 10000
52              DBIB    : 10000
53              DBMB    : 10000
54              (Other):130000
```

We created two new variables `age` and `ttm` from the dates. The numerical and categorical parts of the data are contained in two different data frames. We also convert categorical variables to integers so that we can save both the numerical and categorical variables to a matrix.

From the inforce, we can get the boundaries of the design points as follows:

```
 1 > L <- apply(datN, 2, min)
 2 > H <- apply(datN, 2, max)
 3 > A <- sapply(lapply(datC, levels), length)
 4 > round(L,2)
 5        gbAmt gmwbBalance   withdrawal    FundValue1
 6     50001.72        0.00         0.00          0.00
 7   FundValue2  FundValue3   FundValue4    FundValue5
 8         0.00        0.00         0.00          0.00
 9   FundValue6  FundValue7   FundValue8    FundValue9
10         0.00        0.00         0.00          0.00
11 FundValue10         age          ttm
12         0.00       34.52         0.59
13 > round(H,2)
14        gbAmt gmwbBalance   withdrawal    FundValue1
15    989204.53   499708.73    499585.73     921548.70
16   FundValue2  FundValue3   FundValue4    FundValue5
17    844322.70   580753.42    483936.90     494381.61
18   FundValue6  FundValue7   FundValue8    FundValue9
19    872706.64   634819.08    562485.37     663196.22
20 FundValue10         age          ttm
21    599675.34       64.46        28.52
22 > A
```

```
23        gender productType
24             2           19
```

In the above code, we used the R function apply to obtain the minimum and
the maximum values of the numerical variables. For example, the variable age
ranges from 34.52 to 64.46. We used the functions levels, lapply, sapply,
and length to get the number of categories of each categorical variable. From
the output, we see that the variable productType has 19 levels or categories.

After converting the categorical variables to dummy variables, the inforce
dataset has 34 dimensions. As mentioned in the previous section, we need to
produce 340 design points. To do that, we proceed as follows:

```
 1 > k <- 340
 2 > nSamples <- 50
 3 > vScore <- c()
 4 > maxScore <- 0
 5 > bestS <- NULL
 6 > set.seed(123)
 7 > for(i in 1:nSamples) {
 8 +    mS <- mylhs(k,L,H,A)
 9 +    mD <- calDist(mS, L, H)
10 +    score <- calScore(mD)
11 +    if(score > maxScore) {
12 +       maxScore <- score
13 +       bestS <- mS
14 +    }
15 +    vScore <- c(vScore, score)
16 + }
17 > vScore
18  [1]  2.132743 2.188791 2.156342 2.324484 2.050147
19  [6]  2.536873 2.675516 2.253687 2.398230 2.533923
20 [11]  2.076696 2.563422 2.294985 2.176991 2.654867
21 [16]  2.280236 2.044248 1.932153 2.035398 2.312684
22 [21]  2.418879 2.348083 2.138643 2.238938 2.100295
23 [26]  2.218289 1.994100 2.587021 1.781711 2.088496
24 [31]  2.159292 2.023599 2.643068 2.551622 2.398230
25 [36]  2.654867 1.961652 2.297935 2.545723 2.410029
26 [41]  2.103245 2.271386 2.486726 2.404130 2.365782
27 [46]  2.356932 2.480826 2.489676 2.634218 1.283186
28 > maxScore
29 [1] 2.675516
30 > head(bestS)
31           [,1]       [,2]       [,3]       [,4]       [,5]
32 [1,]  415709.0 187206.51 487796.09 693200.4 699866.3
33 [2,]  914400.8 247643.26 374320.87 720384.7 151928.3
34 [3,]  243937.4 190154.65 316846.41 432230.8 455784.8
35 [4,]  238396.4 368516.76  26526.68 184853.4 590278.7
36 [5,]  313200.1  78125.55 336004.56 695918.8 687413.2
37 [6,]  210691.3 107606.89 372847.17 394172.7 119550.1
```

```
38               [,6]         [,7]         [,8]          [,9]
39 [1,] 260396.81 149891.96 450631.03   10297.42
40 [2,]  71951.75 191290.69 485631.50 218820.25
41 [3,] 541351.27 195573.32 110834.82 321794.48
42 [4,] 428284.23 476799.19 417088.91 723394.00
43 [5,] 467686.38 147036.88  78751.05 432491.79
44 [6,] 459120.70  59956.78 112293.17 399025.16
45              [,10]        [,11]        [,12]         [,13]      [,14]
46 [1,]  395123.4 252205.83 528209.4 566065.21 39.37988
47 [2,]  507480.7  41481.22 491039.1 468772.76 63.83942
48 [3,]  174153.9 506070.91 369746.6 532455.09 61.80849
49 [4,]  342689.9 348442.26 494951.8 122057.81 35.93611
50 [5,]  430703.2  49777.47 283668.0  26534.31 59.77755
51 [6,]  522461.7 242250.34  31301.3 534224.05 37.70215
52            [,15] [,16] [,17]
53 [1,] 16.32507     1    11
54 [2,] 28.02614     1     9
55 [3,] 23.08202     1     7
56 [4,] 15.41864     2    17
57 [5,] 16.57227     2    12
58 [6,] 21.76359     1    18
```

In the above code, we generated 50 Latin hypercube samples and selected the
one with the highest score. From the output, we see that the worst sample
has a score of 1.283186. The best sample has a score of 2.675516.

Now we have obtained a Latin hypercube sample that has a relatively
high score. The design points in the sample are not true policies. To get
representative policies, we need to find policies that are nearest to these design
points. We use the following function to find nearest policies:

```
 1 findPolicy <- function(mS, dat, L, H) {
 2    ind <- c()
 3    k <- nrow(mS)
 4    n <- nrow(dat)
 5    d1 <- length(L)
 6    d2 <- ncol(dat) - d1
 7
 8    for(i in 1:k) {
 9      mN <- dat[,1:d1]      # continuous part
10      mC <- dat[,-(1:d1)]   # categorical part
11
12      vD <- abs(mN - matrix(mS[i,1:d1], nrow=n, ncol=d1,
            byrow=T)) %*%
13        matrix( 1/(H-L), ncol=1) +
14        apply(sign(abs( mC - matrix(mS[i,-(1:d1)], nrow=n,
              ncol=d2, byrow=T) )), 1, sum)
15      tmp <- setdiff(order(vD), ind)
16      ind <- c(ind, tmp[1])
17    }
```

```
18
19    return(sort(ind))
20 }
```

The function has four arguments. The first argument is a Latin hypercube sample. The second argument is a matrix representing an inforce. In the above function, we calculate the distance between each design point and each policy. Then we select a policy that is closest to each design point. If a policy is already selected, we select the near available. Finally, the function returns a sorted vector of indices of the selected representative policies.

Using the above function, we can get 340 representative policies as follows:

```
1 > {
2 + t1 <- proc.time()
3 + lhs <- findPolicy(bestS, datM, L, H)
4 + proc.time() - t1
5 + }
6    user  system elapsed
7  162.07    3.59  166.84
8 > lhs[1:50]
9   [1]  1036  1978  2208  2464  2884  3520  4233  4742
10  [9]  5489  6999  7371  7582  7666  7945  8052  8319
11 [17]  9042  9072  9154  9232  9938 11225 11600 11662
12 [25] 12165 12263 13523 13886 14459 14534 16082 16577
13 [33] 18155 18284 19050 19924 21064 21419 21951 23243
14 [41] 23302 23321 23911 23966 25192 28475 29545 31596
15 [49] 32480 33406
16 > summary(datN[lhs,])
17      gbAmt          gmwbBalance          withdrawal
18  Min.   :103883  Min.   :      0   Min.   :      0
19  1st Qu.:463100  1st Qu.:      0   1st Qu.:      0
20  Median :495839  Median :      0   Median :      0
21  Mean   :515941  Mean   :  66587   Mean   :  25578
22  3rd Qu.:573611  3rd Qu.:      0   3rd Qu.:      0
23  Max.   :897114  Max.   :493348   Max.   :411163
24    FundValue1       FundValue2
25  Min.   :     0  Min.   :      0
26  1st Qu.:     0  1st Qu.:      0
27  Median :     0  Median :      0
28  Mean   : 51317  Mean   :  55846
29  3rd Qu.: 87046  3rd Qu.:  85436
30  Max.   :731605  Max.   :813886
31    FundValue3        FundValue4
32  Min.   :    0.0  Min.   :      0
33  1st Qu.:    0.0  1st Qu.:      0
34  Median :  698.5  Median :      0
35  Mean   :44267.4  Mean   :  27092
36  3rd Qu.:65531.2  3rd Qu.:  38801
37  Max.   :454508.2  Max.   :483937
```

```
38      FundValue5            FundValue6              FundValue7
39   Min.   :      0     Min.   :      0      Min.   :       0
40   1st Qu.:      0     1st Qu.:      0      1st Qu.:       0
41   Median :  36193     Median :      0      Median :   54630
42   Mean   :  51766     Mean   :  58024      Mean   :   61740
43   3rd Qu.:  60956     3rd Qu.:  92414      3rd Qu.:   84202
44   Max.   : 475695     Max.   : 872707      Max.   :  581631
45      FundValue8            FundValue9             FundValue10
46   Min.   :      0     Min.   :      0      Min.   :       0
47   1st Qu.:      0     1st Qu.:      0      1st Qu.:       0
48   Median :  12770     Median :      0      Median :       0
49   Mean   :  45839     Mean   :  42888      Mean   :   43603
50   3rd Qu.:  65776     3rd Qu.:  66370      3rd Qu.:   64319
51   Max.   : 498654     Max.   : 551405      Max.   :  466824
52         age                 ttm
53   Min.   :34.52     Min.   : 2.085
54   1st Qu.:42.34     1st Qu.:10.658
55   Median :49.45     Median :14.763
56   Mean   :49.53     Mean   :15.567
57   3rd Qu.:56.98     3rd Qu.:21.600
58   Max.   :64.37     Max.   :28.353
59 > table(datC[lhs,])
60        productType
61 gender ABRP ABRU ABSU DBAB DBIB DBMB DBRP DBRU DBSU
62      F   13   13    3   10    9    8   14    5   13
63      M    8    8    6    9   10    7    7    8    9
64        productType
65 gender DBWB IBRP IBRU IBSU MBRP MBRU MBSU WBRP WBRU
66      F   14   10    8    9    4   12   13    5   14
67      M   11    7    8   11    7    3   11   14    9
68        productType
69 gender WBSU
70      F    7
71      M    3
```

From the output, we see that it took some time to obtain the representative
policies due to the large number of distance calculations. The summary statistics of the continuous variables show that the representative policies indeed
fill the space well. All combinations of the gender and the product type are
covered.

To see how the representative policies fill the space, we can create a scatter
plot matrix as follows:

```
1 pairs(datN[lhs,])
```

The resulting scatter plot matrix is shown in Figure 4.3. From the figure, we
see that the representative policies fill well the subspace formed by ttm and

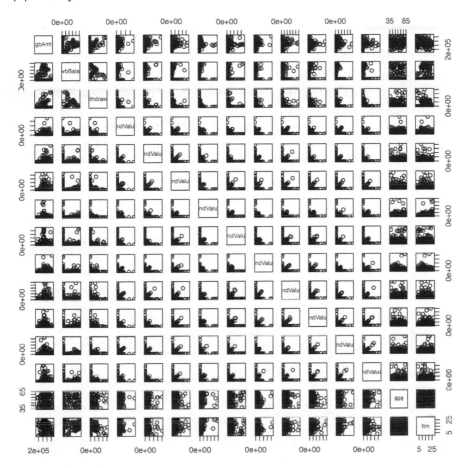

FIGURE 4.3: A scatter plot matrix of the numerical variables based on the representative policies selected by the modified Latin hypercube sampling.

age. Since other variables are highly skewed, the representative policies do not fill the whole subspace formed by these variables.

We can save the indices of the representative policies to a file so that we can use them in future analysis. To save the indices to a CSV file, we use the following code:

```
1 > write.table(lhs, "lhs.csv", sep=",", quote=F, col.
    names=F)
```

After executing the above code, the indices are saved to the file `lhs.csv` in the working directory.

4.4 Summary

In this chapter, we implemented a modified Latin hypercube sampling (LHS) method in R that can handle categorical variables. We also applied the method to a synthetic variable annuity dataset to select representative policies. To make the R program efficient, we used the vectorization trick in many places. For more information about LHS, readers are referred to McKay et al. (1979), McKay and Wanless (2008), Pistone and Vicario (2010), Petelet et al. (2010), and Viana (2013).

5

Conditional Latin Hypercube Sampling

In Chapter 4, we introduced Latin hypercube sampling, which is also referred to as unconditional Latin hypercube sampling. In Latin hypercube sampling, we first create Latin hypercube samples from a space and then find representative policies from an inforce that are closest to the design points. In this chapter, we introduce conditional Latin hypercube sampling, which can be used to select representative policies from an inforce directly. With conditional Latin hypercube sampling, the empirical distributions of the input features are preserved by the selected sample.

5.1 Description of the Method

In conditional Latin hypercube sampling (cLHS), a subset is selected from a dataset in such a way that the subset forms a Latin hypercube (Minasny and McBratney, 2006; Roudier, 2011). To select a subset of k points from a dataset of n points, there are

$$\binom{n}{k} = \frac{n!}{k!(n-k)!}$$

ways. When both n and k are large, the number of combinations is huge. Minasny and McBratney (2006) proposed a search algorithm based on heuristic rules to find conditional Latin hypercubes. The algorithm has been implemented in the R package `clhs` (Roudier, 2011). In what follows, we describe how to use conditional Latin hypercube sampling to select representative variable annuity policies.

Let $X = \{\mathbf{x}_1, \mathbf{x}_2, \ldots, \mathbf{x}_n\}$ be an inforce of n variable annuity policies, each of which is characterized by d attributes or variables. Without loss of generality, suppose that the first d_1 variables are continuous and the remaining $d_2 = d - d_1$ variables are categorical. The search algorithm used to select k representative policies is described as follows.

First, the quantiles of each continuous variable are divided into k strata. Let the quantiles of a continuous variable be q_j^i for $j = 1, 2, \ldots, d_1$ and $i = 1, 2, \cdots, k+1$. Second, k policies $Z = \{\mathbf{z}_1, \mathbf{z}_2, \ldots, \mathbf{z}_k\}$ are selected from X and the following objective function is calculated:

$$O = w_1 O_1 + w_2 O_2 + w_3 O_3, \tag{5.1}$$

where w_1, w_2 and w_3 are the weights of the three components O_1, O_2, and O_3, respectively.

The first component O_1 is the objective function for continuous variables:

$$O_1 = \sum_{j=1}^{d_1} \sum_{i=1}^{n} |\eta_j^i - 1|, \tag{5.2}$$

where η_j^i is the number of z_{ij}, the value of the jth variable of $\mathbf{z}_i$, that fall in between quantiles q_j^i and q_j^{i+1}. This component of the objective function is used to match the empirical distributions of the continuous variables.

The second component O_2 is the objective function for categorical variables:

$$O_2 = \sum_{j=d_1+1}^{d} \sum_{i=1}^{c_j} \left| \frac{\eta_j^i}{k} - \kappa_j^i \right|, \tag{5.3}$$

where c_j is the number of levels or categories of the jth variable, η_j^i is the number of z_{ij} that belong to the ith category of the jth variable, and κ_j^i is the proportion of X in the ith category of the jth variable. This component is used to match the empirical distributions of the categorical variables.

The third component O_3 is used to ensure the correlation of the sampled continuous variables will replicate the original data and is defined as

$$O_3 = \sum_{j=1}^{d_1} \sum_{l=1}^{d_1} |c_{jl} - t_{jl}|, \tag{5.4}$$

where c_{jl} and t_{jl} are the elements of the sample correlation matrix of the continuous data of X and the sample correlation matrix of the continuous data of Z, respectively. This component is used to preserve correlations.

An annealing schedule (i.e., the simulated annealing algorithm) is used to obtain the final sample. By construction, the final sample obtained from the search algorithm preserves the empirical distribution and multivariate correlation of the original data.

5.2 Implementation

We will use the R package `clhs` for conditional Latin hypercube sampling. If this package is not yet installed, you can install it by executing the following command:

```
1 install.packages("clhs")
```

Once the package is installed, we can load it into R by running the following command:

```
1 library(clhs)
```

The function used to create a conditional Latin hypercube sample is `clhs`. The following piece of code contains a simple illustration of how to use this function:

```
1 > set.seed(1)
2 > df <- data.frame(a = runif(1000), b = rnorm(1000))
3 >
4 > set.seed(1)
5 > res <- clhs(df, size = 50, iter = 100, simple = T)
6   |=========================================| 100%
7 > res
8  [1] 266 372 572 906 940 380 707 638 478 754 436 256
9 [13] 767 142  14 934 356 458 758 285 904 395 109 220
10 [25] 165 260 353 403 938 133 251 642 625 700 377 843
11 [37] 400 434  82 872 831 149  11   2 745 890   3 307
12 [49] 547 896
```

In the above code, we first created a dataset with two variables, a uniformly distributed variable and a normally distributed variable. Then we call the function `clhs` to generate a Latin hypercube sample from the dataset. Since we set the argument `simple` to be true, the output from the function is a list of indices of the selected points. We set the random seeds so that we can replicate our results.

Since the dataset is two-dimensional, we can visualize the dataset and the selected sample as follows:

```
1 > plot(df$a,df$b, xlim=range(df$a), ylim=range(df$b))
2 > plot(df$a[res],df$b[res], xlim=range(df$a), ylim=range
      (df$b))
3 > plot(df$a[res],df$b[res], xlim=range(df$a), ylim=range
      (df$b))
```

The resulting plots are shown in Figure 5.1. From the figure, we can see that the conditional Latin hypercube sample is a subset of the original dataset.

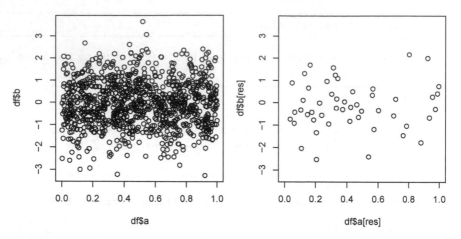

FIGURE 5.1: A two-dimensional dataset (left) and a conditional Latin hypercube sample produced by `clhs` (right).

5.3 Examples

In this section, we apply the function `clhs` to select representative variable annuity policies from the synthetic inforce described in Section A.1.

First, we load the inforce into R. Suppose that the inforce is saved in the current working directory. Then we load the dataset as follows:

```
 1 > inforce <- read.csv("inforce.csv")
 2 >
 3 > vNames <- c("gbAmt", "gmwbBalance", "withdrawal",
       paste("FundValue", 1:10, sep=""))
 4 >
 5 > age <- with(inforce, (currentDate-birthDate)/365)
 6 > ttm <- with(inforce, (matDate - currentDate)/365)
 7 >
 8 > datN <- cbind(inforce[,vNames], data.frame(age=age,
       ttm=ttm))
 9 > datC <- inforce[,c("gender", "productType")]
10 > dat <- cbind(datN, datC)
11 > head(dat)
12       gbAmt gmwbBalance withdrawal FundValue1
13 1  87657.37          0          0       0.00
14 2 161534.10          0          0   25681.18
15 3 407190.05          0          0       0.00
16 4 307425.14          0          0       0.00
17 5 356480.86          0          0       0.00
18 6 275079.73          0          0   34449.27
```

19	FundValue2	FundValue3	FundValue4	FundValue5
20 1	0.00	0.00	45008.86	0.00
21 2	0.00	23096.98	16719.40	19297.09
22 3	0.00	0.00	0.00	389147.88
23 4	0.00	52957.63	0.00	85110.56
24 5	27505.37	12784.85	14425.32	31306.97
25 6	33844.28	31926.71	0.00	29442.19
26	FundValue6	FundValue7	FundValue8	FundValue9
27 1	0.00	0.00	0.00	0.00
28 2	25791.35	24386.58	0.00	23949.69
29 3	0.00	0.00	0.00	0.00
30 4	0.00	0.00	0.00	62465.01
31 5	24655.37	17119.72	18851.82	16424.64
32 6	34228.91	33179.90	30762.85	32515.50

33	FundValue10	age	ttm	gender	productType
34 1	0.00	47.36164	19.26575	F	ABRP
35 2	21906.86	57.78630	18.18082	M	ABRP
36 3	0.00	53.12055	18.18082	M	ABRP
37 4	0.00	42.86301	14.68219	M	ABRP
38 5	21829.42	62.70959	11.17534	F	ABRP
39 6	31363.51	52.70137	19.92877	M	ABRP

In the above code, we also created a new data frame called `dat` from the original inforce by selecting relevant variables.

Similar as done in Chapter 4, we want to select 340 representative policies. To do that, we apply the function `clhs` as follows:

```
 1 > set.seed(1)
 2 > {
 3 + t1 <- proc.time()
 4 + res <- clhs(dat, size = 340, iter = 1000, simple = T)
 5 + proc.time() - t1
 6 + }
 7 |===========================================| 100%
 8    user  system elapsed
 9   38.24    1.02   39.68
10 > res <- sort(res)
11 > res[1:50]
12  [1]    383   728  1276  1871  1901  2484  4082  4578
13  [9]   5272  5855  7684  8802  9174  9569  9612  9899
14 [17]  11959 12230 12984 13436 13790 13909 14291 14412
15 [25]  14633 14634 14872 15246 15766 17262 17263 17995
16 [33]  18767 18813 19006 20174 20646 20947 21251 22230
17 [41]  22557 22821 23346 24368 24514 24756 24931 25086
18 [49]  25477 25538
```

In the above code, we used the function `clhs` to select 340 representative policies from the inforce. The number of iterations was set to 1,000. Under this setting, it took the function about 40 seconds to finish. The indices of the first 50 representative policies arc also shown above.

Some summary statistics of the sample are produced below:

```
 1 > summary(datN[res,])
 2      gbAmt             gmwbBalance            withdrawal
 3  Min.   : 53935    Min.   :      0    Min.   :      0
 4  1st Qu.:195678    1st Qu.:      0    1st Qu.:      0
 5  Median :317622    Median :      0    Median :      0
 6  Mean   :320547    Mean   :  39440    Mean   :  21242
 7  3rd Qu.:445560    3rd Qu.:      0    3rd Qu.:      0
 8  Max.   :844045    Max.   :431026    Max.   :361183
 9    FundValue1          FundValue2           FundValue3
10  Min.   :      0    Min.   :      0    Min.   :      0
11  1st Qu.:      0    1st Qu.:      0    1st Qu.:      0
12  Median :  14082    Median :  17014    Median :   8641
13  Mean   :  27804    Mean   :  26564    Mean   :  18486
14  3rd Qu.:  40007    3rd Qu.:  39876    3rd Qu.:  25876
15  Max.   :487435    Max.   :278646    Max.   :361023
16    FundValue4          FundValue5           FundValue6
17  Min.   :      0    Min.   :      0    Min.   :      0
18  1st Qu.:      0    1st Qu.:      0    1st Qu.:      0
19  Median :   8369    Median :  16260    Median :  16786
20  Mean   :  15061    Mean   :  25258    Mean   :  28855
21  3rd Qu.:  23354    3rd Qu.:  37619    3rd Qu.:  43026
22  Max.   :178019    Max.   :448109    Max.   :250540
23    FundValue7          FundValue8           FundValue9
24  Min.   :      0    Min.   :      0    Min.   :      0
25  1st Qu.:      0    1st Qu.:      0    1st Qu.:      0
26  Median :  11386    Median :  11158    Median :  10840
27  Mean   :  21099    Mean   :  21347    Mean   :  21482
28  3rd Qu.:  31060    3rd Qu.:  31053    3rd Qu.:  30160
29  Max.   :237332    Max.   :320751    Max.   :551405
30   FundValue10            age                  ttm
31  Min.   :      0    Min.   :34.52    Min.   : 0.6712
32  1st Qu.:      0    1st Qu.:41.44    1st Qu.:10.7370
33  Median :  12435    Median :49.78    Median :15.2630
34  Mean   :  20805    Mean   :49.07    Mean   :15.1455
35  3rd Qu.:  32862    3rd Qu.:55.98    3rd Qu.:19.9288
36  Max.   :291362    Max.   :64.46    Max.   :28.0192
37 > table(datC[res,])
38        productType
39 gender ABRP ABRU ABSU DBAB DBIB DBMB DBRP DBRU DBSU
40      F    7    5    7    5    4    7    7    3    4
41      M    9   14   10   11   14    8   13    7   14
42        productType
43 gender DBWB IBRP IBRU IBSU MBRP MBRU MBSU WBRP WBRU
44      F    8   10    6    6    8   13    6    7    9
45      M    8    8   14   11   12    6   10   15   16
```

```
46              productType
47 gender  WBSU
48       F      8
49       M     10
```

By comparing the summary statistics shown above with those given in Appendix A, we see that the representative policies selected by conditional Latin hypercube sampling fill the inforce space well. In addition, all combinations of the gender and the product type are covered.

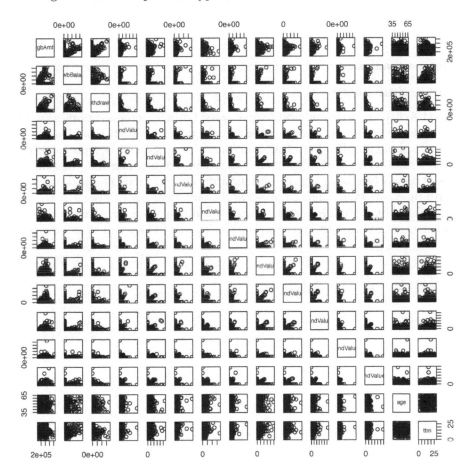

FIGURE 5.2: A scatter plot matrix of the numerical variables based on the representative policies selected by conditional Latin hypercube sampling.

To visualize the representative policies, we create a scatter plot matrix of the numerical variables as follows:

```
1 > pairs(datN[res,])
```

The resulting plot is shown in Figure 5.2. We see similar patterns as in Figure 4.3. For variables `age` and `ttm`, the representative policies fill the space well. Since other variables are skewed, we see that the representative policies do not fill the spaces formed by these variables.

To retain the indices of the representative policies for future analysis, we save the indices to a CSV file named `clhs.csv` as follows:

```
1 > write.table(res, "clhs.csv", sep=",", quote=F, col.
    names=F)
```

5.4 Summary

In this chapter, we introduced a conditional Latin hypercube sampling method for selecting representative variable annuity policies. Conditional Latin hypercube sampling differs from (unconditional) Latin hypercube sampling in that the former selects design points in the presence of auxiliary information. This is accomplished by using an optimization algorithm to minimize an objective function, which aims to preserve the empirical distributions of the input features. For more information about conditional Latin hypercube sampling, readers are referred to Minasny and McBratney (2006) and Roudier (2011).

6

Hierarchical k-Means

In Chapters 4 and 5, we described how to select representative policies that form a Latin hypercube. In this chapter, we introduce an approach based on data clustering, which is a form of unsupervised learning and refers to the process of dividing a dataset into homogeneous groups or clusters. In particular, we introduce hierarchical k-means, which is efficient in dividing a large dataset into many clusters.

6.1 Description of the Method

The hierarchical k-means is an extension of the well-known k-means algorithm. The algorithm was proposed by Nister and Stewenius (2006). Gan and Valdez (2018a) adopted this algorithm for selecting representative variable annuity policies.

To introduce the hierarchical k-means algorithm, let us first introduce the k-means algorithm, which was independently developed by Sebestyen (1962) and MacQueen (1967) as a strategy that attempts to minimize within-group variation.

Let $X = \{\mathbf{x}_1, \mathbf{x}_2, \ldots, \mathbf{x}_n\}$ be a set of n data points, each of which is described by a numerical vector. The goal of the k-means algorithm is to divide the dataset into k clusters by minimizing the following objective function:

$$P(U, Z) = \sum_{l=1}^{k} \sum_{i=1}^{n} u_{il} \|\mathbf{x}_i - \mathbf{z}_l\|^2, \qquad (6.1)$$

where k is the desired number of clusters specified by the user, $U = (u_{il})_{n \times k}$ is an $n \times k$ partition matrix, $Z = \{\mathbf{z}_1, \mathbf{z}_2, \ldots, \mathbf{z}_k\}$ is a set of cluster centers, and $\| \cdot \|$ is the L^2 norm or Euclidean distance, i.e.,

$$\|\mathbf{x}_i - \mathbf{z}_l\| = \left(\sum_{j=1}^{d} (x_{ij} - z_{lj})^2 \right)^{\frac{1}{2}}. \qquad (6.2)$$

Here d is the number of dimensions of the data, x_{ij} denotes the jth component of the vector $\mathbf{x}_i$, and z_{lj} denotes the jth component of the vector $\mathbf{z}_l$.

The partition matrix U is a binary matrix that contains the information about the cluster memberships of the individual data points. The (i, l)-th entry of the partition matrix can be interpreted as follows:

$$u_{il} = \begin{cases} 1, & \text{if } \mathbf{x}_i \text{ belongs to the } j\text{th cluster,} \\ 0, & \text{if otherwise.} \end{cases}$$

Hence it satisfies the following conditions:

$$u_{il} \in \{0, 1\}, \quad i = 1, 2, \ldots, n, \ l = 1, 2, \ldots, k, \tag{6.3a}$$

$$\sum_{l=1}^{k} u_{il} = 1, \quad i = 1, 2, \ldots, n. \tag{6.3b}$$

The condition given in Equation (6.3b) means that a data point can belong to one and only one cluster.

The k-means algorithm is an approximate algorithm that aims to minimize the objective function given in Equation (6.1). It consists of the following two phases:

The initialization phase In the initialization phase, k initial cluster centers are selected randomly from the input dataset.

The iteration phase In the iteration phase, the algorithm alternatively updates the partition matrix U by fixing Z and updates the cluster centers Z by fixing U until some criterion (e.g., maximum number of iterations) is met. Theorem 6.1 and Theorem 6.2 specify how the updating is done. See Gan (2011).

Theorem 6.1. *Let the cluster centers $Z = \{\mathbf{z}_1, \mathbf{z}_2, \ldots, \mathbf{z}_k\}$ be fixed. Then the objective function given in Equation (6.1) is minimized if and only if*

$$u_{il} = \begin{cases} 1, & \text{if } \|\mathbf{x}_i - \mathbf{z}_l\| = \min_{1 \le j \le k} \|\mathbf{x}_i - \mathbf{z}_j\|; \\ 0, & \text{if otherwise,} \end{cases}$$

for $i = 1, 2, \ldots, n$ and $l = 1, 2, \ldots, k$.

Theorem 6.2. *Let the partition matrix U be fixed. Then the objective function given in Equation (6.1) is minimized if and only if*

$$z_{lj} = \frac{\sum_{i=1}^{n} u_{il} x_{ij}}{\sum_{i=1}^{n} u_{il}}, \quad l = 1, 2, \ldots, k, \ j = 1, 2, \ldots, d.$$

The k-means algorithm is typically implemented as follows:

1. Initialize $\mathbf{z}_1, \mathbf{z}_2, \ldots, \mathbf{z}_k$ by randomly selecting k points from X.
2. Calculate the distance between $\mathbf{x}_i$ and $\mathbf{z}_j$ for all $1 \le i \le n$ and $1 \le j \le k$.

3. Update the partition matrix U according to Theorem 6.1.

4. Update cluster centers Z according to Theorem 6.2.

5. Repeat Steps 2–4 until there are no further changes in the partition matrix.

6. Return the partition matrix U and the cluster centers Z.

From the above procedure, we see that the k-means algorithm alternatively updates the partition matrix and the cluster centers in the iteration phase. Note that there are other criteria to terminate the algorithm. For example, the algorithm can be terminated if the change of the objective function is less than a threshold or a maximum number of iterations is reached.

One drawback of the k-means algorithm is that it is not efficient when used to divide a large dataset into a moderate number (e.g., 100, 200) of clusters (Gan et al., 2016). Hierarchical k-means addresses this scalability issue by using a divisive approach to apply the k-means with small k's repeatedly until the desired number of clusters is reached.

In this section, we use the following hierarchical k-means algorithm adopted by Gan and Valdez (2018a):

1. Apply the k-means algorithm to divide the dataset into two clusters.

2. Apply the k-means algorithm to divide the largest existing cluster into two clusters.

3. Repeat Step 2 until the number of clusters is equal to k.

In this hierarchical k-means algorithm, we use binary splitting to create a hierarchy of clusters.

6.2 Implementation

To implement the hierarchical k-means algorithm described in the previous section, we need a mechanism for storing the tree structure created by hierarchical k-means. It is tedious to implement such a mechanism in R by ourselves. Instead, we will use the R package **data.tree** for this purpose.

If the package **data.tree** has not been installed in your computer, you can execute the following command to install the package:

```
1 install.packages("data.tree")
```

Once the package is loaded into R, it provides an environment called **Node**. We can see the details as follows:

```
1 > mode(Node)
2 [1] "environment"
```

```
 3 > Node
 4 <Node> object generator
 5   Public:
 6     initialize: function (name, check = c("check", "no-
          warn", "no-check"), ...)
 7     AddChild: function (name, check = c("check", "no-
          warn", "no-check"), ...)
 8     AddChildNode: function (child)
 9     AddSibling: function (name, check = c("check", "no-
          warn", "no-check"), ...)
10     AddSiblingNode: function (sibling)
11     RemoveChild: function (name)
12     RemoveAttribute: function (name, stopIfNotAvailable
          = TRUE)
13     Sort: function (attribute, ..., decreasing = FALSE,
          recursive = TRUE)
14     Revert: function (recursive = TRUE)
15     Prune: function (pruneFun)
16     Climb: function (...)
17     Navigate: function (path)
18     FindNode: function (name)
19     Get: function (attribute, ..., traversal = c("pre-
          order", "post-order",
20     Do: function (fun, ..., traversal = c("pre-order", "
          post-order",
21     Set: function (..., traversal = c("pre-order", "post
          -order", "in-order",
22     clone: function (deep = FALSE)
23   Active bindings:
24     name: function (value)
25     parent: function (value)
26     children: function (value)
27     isLeaf: function ()
28     isRoot: function ()
29     count: function ()
30     totalCount: function ()
31     path: function ()
32     pathString: function ()
33     position: function ()
34     fields: function ()
35     fieldsAll: function ()
36     levelName: function ()
37     leaves: function ()
38     leafCount: function ()
39     level: function ()
40     height: function ()
41     isBinary: function ()
42     root: function ()
43     siblings: function ()
```

```
44        averageBranchingFactor: function ()
45    Private:
46      p_name:
47      p_children: NULL
48      p_parent: NULL
49    Parent env: <environment: namespace:data.tree>
50    Locked objects: FALSE
51    Locked class: TRUE
52    Portable: TRUE
```

As we can see from the above output, the environment Node provides many
functions that are useful for tree data structures.

With the help of the package data.tree, we can implement the hierarchical
k-means algorithm as follows:

```
1  require(data.tree)
2  hkmeans <- function(X, k) {
3    res <- Node$new("Node 0")
4    nCount <- 0
5    tmp <- kmeans(X, 2)
6    for(i in 1:2) {
7      nCount <- nCount + 1
8      nodeA <- res$AddChild(paste("Node", nCount))
9      nodeA$members <- names(which(tmp$cluster==i))
10     nodeA$size <- length(nodeA$members)
11     nodeA$center <- tmp$centers[i,]
12   }
13
14   while(TRUE) {
15     vSize <- res$Get("size", filterFun = isLeaf)
16     if(length(vSize) >= k) {
17       break
18     }
19     maxc <- which(vSize == max(vSize))
20     nodeL <- FindNode(res, names(maxc))
21     tmp <- kmeans(X[nodeL$members,], 2)
22     for(i in 1:2) {
23       nCount <- nCount + 1
24       nodeA <- nodeL$AddChild(paste("Node", nCount))
25       nodeA$members <- names(which(tmp$cluster==i))
26       nodeA$size <- length(nodeA$members)
27       nodeA$center <- tmp$centers[i,]
28     }
29   }
30   return(res)
31 }
```

Let us look at the R code shown above. The name of our function is called
hkmeans. At the beginning, we create a root node named Node 0 by using the

environment `Node` provided by the package `data.tree`. In Line 5, we divide the input dataset into two clusters. In Lines 6–12, we save the clustering results into two child nodes of the root node. In Lines 14–29, we repeat dividing the largest existing cluster into two clusters and saving the results to the child nodes of the node that contains the cluster. Here we use an infinite loop to do the binary splitting. Once the desired number of clusters is reached, the infinite loop is terminated by the command `break`. Finally, the function returns the resulting tree structure, which can be further processed to find the representative policies.

6.3   Examples

In this section, we apply the hierarchical *k*-means algorithm to the synthetic inforce described in Section A.1.

First, we load the inforce and select relevant variables by running the following code:

```
1  inforce <- read.csv("inforce.csv")
2
3  vNames <- c("gbAmt", "gmwbBalance", "withdrawal",    paste
        ("FundValue", 1:10, sep=""))
4
5  age <- with(inforce, (currentDate-birthDate)/365)
6  ttm <- with(inforce, (matDate - currentDate)/365)
7
8  datN <- cbind(inforce[,vNames], data.frame(age=age, ttm=
        ttm))
9  datC <- inforce[,c("gender", "productType")]
10 dat <- cbind(datN, datC)
```

Since the hierarchical *k*-means can only handle numerical data, we need to convert categorical variables to vectors of numerical values. To do that, we use the one-hot encoding, which converts a categorical value to a vector of binary numbers such that the component corresponding to the position of the categorical value is one and all other components are zero. To avoid that a single variable dominates the distance, we also need to normalize the variables so that they have the same scales.

To convert categorical variables to vectors of numerical values, we use the function `model.matrix` as follows:

```
1 > X <- model.matrix( ~ ., data=dat)[,-1]
2 > colnames(X)
3  [1] "gbAmt"              "gmwbBalance"
4  [3] "withdrawal"         "FundValue1"
```

```
 5    [5]  "FundValue2"       "FundValue3"
 6    [7]  "FundValue4"       "FundValue5"
 7    [9]  "FundValue6"       "FundValue7"
 8   [11]  "FundValue8"       "FundValue9"
 9   [13]  "FundValue10"      "age"
10   [15]  "ttm"              "genderM"
11   [17]  "productTypeABRU"  "productTypeABSU"
12   [19]  "productTypeDBAB"  "productTypeDBIB"
13   [21]  "productTypeDBMB"  "productTypeDBRP"
14   [23]  "productTypeDBRU"  "productTypeDBSU"
15   [25]  "productTypeDBWB"  "productTypeIBRP"
16   [27]  "productTypeIBRU"  "productTypeIBSU"
17   [29]  "productTypeMBRP"  "productTypeMBRU"
18   [31]  "productTypeMBSU"  "productTypeWBRP"
19   [33]  "productTypeWBRU"  "productTypeWBSU"
20   > summary(X[,1:10])
21        gbAmt           gmwbBalance          withdrawal
22   Min.   : 50002   Min.   :      0    Min.   :      0
23   1st Qu.:179759   1st Qu.:      0    1st Qu.:      0
24   Median :303525   Median :      0    Median :      0
25   Mean   :313507   Mean   : 36141    Mean   : 21928
26   3rd Qu.:427544   3rd Qu.:      0    3rd Qu.:      0
27   Max.   :989205   Max.   :499709    Max.   :499586
28      FundValue1        FundValue2         FundValue3
29   Min.   :      0  Min.   :      0    Min.   :      0
30   1st Qu.:      0  1st Qu.:      0    1st Qu.:      0
31   Median :   8299  Median :   8394    Median :   4942
32   Mean   :  26611  Mean   :  26045    Mean   :  17391
33   3rd Qu.:  39209  3rd Qu.:  38463    3rd Qu.:  24251
34   Max.   : 921549  Max.   : 844323    Max.   : 580753
35      FundValue4        FundValue5         FundValue6
36   Min.   :      0  Min.   :      0    Min.   :      0
37   1st Qu.:      0  1st Qu.:      0    1st Qu.:      0
38   Median :   4225  Median :   7248    Median :   8556
39   Mean   :  14507  Mean   :  21041    Mean   :  26570
40   3rd Qu.:  20756  3rd Qu.:  32112    3rd Qu.:  39241
41   Max.   : 483937  Max.   : 494382    Max.   : 872707
42      FundValue7
43   Min.   :      0
44   1st Qu.:      0
45   Median :   6602
46   Mean   :  21506
47   3rd Qu.:  31088
48   Max.   : 634819
```

From the above output, we see that the categorical variable **gender** is replaced by a numerical variable. We also see the summary statistics of the first ten

variables. We see that these variables have huge values, which will dominate
the distance used in data clustering.

To normalize the data, we proceed as follows:

```
 1 > vMin <- apply(X, 2, min)
 2 > vMax <- apply(X, 2, max)
 3 > X <- (X - matrix(vMin, nrow=nrow(X), ncol= ncol(X),
       byrow=TRUE)) / matrix(vMax-vMin, nrow=nrow(X), ncol=
       ncol(X), byrow=TRUE)
 4 > summary(X[,1:10])
 5      gbAmt              gmwbBalance            withdrawal
 6  Min.   :0.0000    Min.   :0.00000    Min.    :0.00000
 7  1st Qu.:0.1382    1st Qu.:0.00000    1st Qu.:0.00000
 8  Median :0.2699    Median :0.00000    Median :0.00000
 9  Mean   :0.2806    Mean   :0.07232    Mean    :0.04389
10  3rd Qu.:0.4020    3rd Qu.:0.00000    3rd Qu.:0.00000
11  Max.   :1.0000    Max.   :1.00000    Max.    :1.00000
12    FundValue1         FundValue2            FundValue3
13  Min.    :0.000000   Min.    :0.000000   Min.    :0.000000
14  1st Qu.:0.000000   1st Qu.:0.000000   1st Qu.:0.000000
15  Median :0.009006   Median :0.009942   Median :0.008509
16  Mean    :0.028877   Mean    :0.030847   Mean    :0.029946
17  3rd Qu.:0.042547   3rd Qu.:0.045555   3rd Qu.:0.041759
18  Max.    :1.000000   Max.    :1.000000   Max.    :1.000000
19    FundValue4         FundValue5            FundValue6
20  Min.    :0.000000   Min.    :0.00000   Min.    :0.000000
21  1st Qu.:0.000000   1st Qu.:0.00000   1st Qu.:0.000000
22  Median :0.008731   Median :0.01466   Median :0.009804
23  Mean    :0.029978   Mean    :0.04256   Mean    :0.030445
24  3rd Qu.:0.042890   3rd Qu.:0.06495   3rd Qu.:0.044965
25  Max.    :1.000000   Max.    :1.00000   Max.    :1.000000
26    FundValue7
27  Min.    :0.00000
28  1st Qu.:0.00000
29  Median :0.01040
30  Mean    :0.03388
31  3rd Qu.:0.04897
32  Max.    :1.00000
```

Here we normalize each variable to the interval $[0, 1]$ by using the minmax
normalization method. From the output, we see that the variables are appropriately normalized.

Now we are ready to apply the hierarchical k-means algorithm to divide
the dataset. Similar to Chapters 4 and 5, we want to select 340 representative
policies. To do that, we divide the dataset into 340 clusters as follows:

```
 1 > {
 2 + set.seed(123)
 3 + t1 <- proc.time()
```

```
4 + res <- hkmeans(X, 340)
5 + proc.time() - t1
6 + }
7    user    system  elapsed
8   12.71     0.58    13.40
```

The output shows that the hierarchical k-means is fairly fast. In this case, it took about 13 seconds for the algorithm to complete. We can visualize the tree structure as follows:

```
1 > par(mar=c(0,4,1,1))
2 > plot(as.dendrogram(res), center = TRUE, leaflab="none"
      )
```

The resulting plot is shown in Figure 6.1, from which we can see how the clusters are nested. We did not plot the node names due to the space constraint.

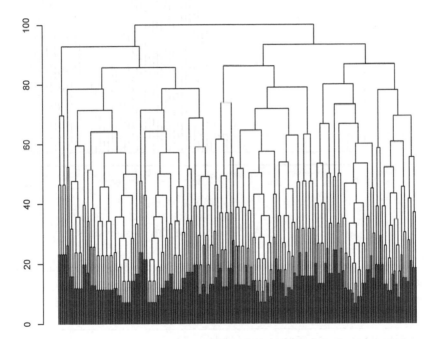

FIGURE 6.1: A plot of the tree structure produced by the hierarchical k-means algorithm.

To see the sizes of terminal clusters, we use the function Get to retrieve the cluster sizes as follows:

```
1 par(mar=c(4,4,1,1))
2 hist(vSize, br=100)
```

We also plotted a histogram of the cluster sizes. The histogram is shown
in Figure 6.2. Since we divide the largest existing cluster, the sizes of most
terminal clusters are relatively close.

FIGURE 6.2: A histogram of the cluster sizes.

Once we get the tree structure that contains the 100 terminal clusters, we
need a way to find the policy that is closest to the center in each terminal
cluster. To do that, we use the function Do to apply the following function to
each leaf node:

```
1 findPolicy <- function(node, X) {
2   z <- node$center
3   vD <- apply( (X[node$members,] - matrix(z, nrow=node$
        size, ncol=length(z), byrow=T))^2, 1, sum)
4   iMin <- which(vD == min(vD))
5   node$policy <- node$members[iMin]
6 }
7 res$Do(findPolicy, filterFun = isLeaf, X=X)
```

Executing the above code will add the index of the policy that is closest to the cluster center to the node.

Then we can extract the indices of the representative policies from the tree structure as follows:

```
1 > vInd <-  res$Get("policy", filterFun = isLeaf)
2 > vInd <- sort(as.numeric(vInd))
3 > print(unname(vInd[1:50]))
4 [1]    302    958  1530  1786  1871  1937  3007  3204
      3619
5 [10]  4201  4228  5877  6525  7115  7712  7896  8147
      8802
6 [19] 11362 11725 11823 11886 12155 12282 12934 13047
      13500
7 [28] 14082 14384 14604 15496 16431 16932 17005 17185
      19603
8 [37] 19968 20642 20893 21541 21570 21674 21810 21929
      22515
9 [46] 22787 22983 24019 24528 24761
```

The following code shows the summary statistics of the representative policies:

```
1 > summary(datN[vInd,])
2       gbAmt            gmwbBalance         withdrawal
3  Min.   :113899   Min.   :      0   Min.   :      0
4  1st Qu.:186755   1st Qu.:      0   1st Qu.:      0
5  Median :303172   Median :      0   Median :      0
6  Mean   :313176   Mean   : 34568   Mean   : 21996
7  3rd Qu.:406316   3rd Qu.:      0   3rd Qu.:      0
8  Max.   :744813   Max.   :374742   Max.   :346577
9    FundValue1        FundValue2        FundValue3
10 Min.   :    0   Min.   :    0   Min.   :    0
11 1st Qu.:14980   1st Qu.:14664   1st Qu.: 8606
12 Median :24204   Median :23938   Median :15105
13 Mean   :25223   Mean   :24584   Mean   :15648
14 3rd Qu.:37871   3rd Qu.:35137   3rd Qu.:22224
15 Max.   :63662   Max.   :59499   Max.   :50219
16   FundValue4        FundValue5        FundValue6
17 Min.   :    0   Min.   :    0   Min.   :     0
18 1st Qu.: 8129   1st Qu.:13125   1st Qu.: 15074
19 Median :12938   Median :20778   Median : 23816
20 Mean   :13917   Mean   :20562   Mean   : 26326
21 3rd Qu.:19779   3rd Qu.:30603   3rd Qu.: 36495
22 Max.   :35443   Max.   :44818   Max.   :450939
23   FundValue7        FundValue8        FundValue9
24 Min.   :    0   Min.   :    0   Min.   :    0
25 1st Qu.:11457   1st Qu.:11515   1st Qu.:10121
26 Median :19581   Median :18382   Median :17641
```

```
27  Mean     :20032    Mean     :18966    Mean     :17988
28  3rd Qu.:29379      3rd Qu.:27865      3rd Qu.:26094
29  Max.     :53004    Max.     :44575    Max.     :52217
30   FundValue10            age                 ttm
31  Min.    :    0    Min.    :37.52    Min.   : 6.088
32  1st Qu.:12137     1st Qu.:42.03     1st Qu.: 9.820
33  Median :19271     Median :47.37     Median :13.468
34  Mean    :19849    Mean    :49.44    Mean    :14.464
35  3rd Qu.:29137     3rd Qu.:56.87     3rd Qu.:19.348
36  Max.    :47302    Max.    :61.71    Max.    :23.184
37  > table(datC[vInd,])
38         productType
39  gender ABRP ABRU ABSU DBAB DBIB DBMB DBRP DBRU DBSU DBWB
40      F    8    8    8    8    8    8    8    8    8    7
41      M   10   11   10    9   10    9    9   13    9   11
42         productType
43  gender IBRP IBRU IBSU MBRP MBRU MBSU WBRP WBRU WBSU
44      F    8    8    8    8    7    8    7    7    6
45      M    8   13    8   10   12   10   11   11   10
```

From the output, we see that the numerical variables do not spread well in the input space. However, all combinations of the categorical variables are covered by the set of representative policies.

Figure 6.3 shows a scatter plot matrix of representative policies. The patterns are slightly different from those shown in Figures 4.3 and 5.2. To save the indices to a CSV file for future analysis, we use the following code:

```
1 > write.table(vInd, "hkmeans.csv", sep=",", quote=F, col
    .names=F)
```

6.4   Summary

Data clustering (also known as cluster analysis) refers to the process of dividing a set of objects into homogeneous groups or clusters (Gan, 2011). Most of the existing clustering algorithms are developed to divide a dataset into a small number of clusters. As a result, these clustering algorithms are not efficient in dividing a large dataset into a moderate or large number of clusters. In this chapter, we introduced the hierarchical k-means algorithm, which is efficient in getting a large number of clusters, for the purpose of selecting representative variable annuity polices. For more information about data clustering and its application in actuarial science, readers are referred to Gan and Valdez (2018a).

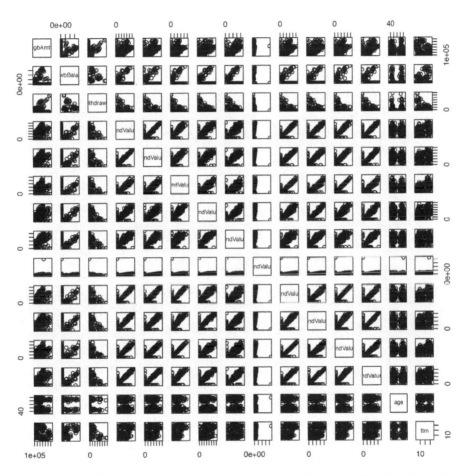

FIGURE 6.3: A scatter plot matrix of the numerical variables based on the representative policies selected by the hierarchical k-means algorithm.

Part III

Metamodels

7

Ordinary Kriging

A major component of a metamodeling approach is the predictive model. In this chapter, we introduce the ordinary kriging method, which can serve as a predictive model. This method can be used to predict the fair market values of the guarantees embedded in variable annuities. Under the ordinary kriging method, the covariance structure of the responses is modeled and the predictions are based on the weighted average of the observed values.

7.1   Description of the Model

Ordinary kriging is a method developed in geostatistics and derives estimates that are often called best linear unbiased estimators (Isaaks and Srivastava, 1990). The estimator is linear because the estimates are weighted linear combinations of the available data. It is unbiased because the mean errors of the estimates are equal to 0. It is also the best because the variance of the errors is minimized.

The ordinary kriging method treats the unknown values as the outcome of a random process and uses a stationary random function to model the values. To describe the kriging method, let V be the stationary random function, $\mathbf{z}_1$, $\mathbf{z}_2$, ..., $\mathbf{z}_k$ be locations (e.g., the representative variable annuity policies) where we a value of the random function is known, and $\mathbf{x}_1$, $\mathbf{x}_2$, ..., $\mathbf{x}_n$ be locations (e.g., policies in the full inforce) where the values of the random function are unknown.

In ordinary kriging, $V(\mathbf{z}_j)$ and $V(\mathbf{x}_i)$ are random variables for all $j = 1, 2, \ldots, k$ and $i = 1, 2, \ldots, n$. For $i = 1, 2, \ldots, n$, the random variable at the unknown location $\mathbf{x}_i$ is estimated as

$$\widehat{V}(\mathbf{x}_i) = \sum_{j=1}^{k} w_{ji} V(\mathbf{z}_j). \tag{7.1}$$

Since the estimate is a weighted linear combination of random variables, it is also a random variable. The estimation error is defined as

$$R(\mathbf{x}_i) = \widehat{V}(\mathbf{x}_i) - V(\mathbf{x}_i). \tag{7.2}$$

69

Combining Equation (7.2) and Equation (7.1), we get

$$R(\mathbf{x}_i) = \sum_{j=1}^{k} w_{ji} V(\mathbf{z}_j) - V(\mathbf{x}_i). \tag{7.3}$$

Taking expectation in both sides of Equation (7.3) and noting that the random function is stationary, we get

$$E[R(\mathbf{x}_i)] = \sum_{j=1}^{k} w_{ji} E[V(\mathbf{z}_j)] - E[V(\mathbf{x}_i)]$$

$$= \sum_{j=1}^{k} w_{ji} E[V] - E[V].$$

To ensure the estimation errors have an expected value of zero, we need the following constraints on the weights:

$$\sum_{j=1}^{k} w_{ji} = 1, \quad i = 1, 2, \ldots, n, \tag{7.4}$$

that is, the weights sum to one.

Now we assume that Equation (7.4) holds. Then we have $E[\widehat{V}(\mathbf{x}_i)] = E[V(\mathbf{x}_i)]$. From Equation (7.2), we obtain the variance of the estimation error as follows:

$$\mathrm{Var}\,(R(\mathbf{x}_i)) = \mathrm{Var}\,(\widehat{V}(\mathbf{x}_i)) - 2\,\mathrm{Cov}\,(\widehat{V}(\mathbf{x}_i), V(\mathbf{x}_i)) + \mathrm{Var}\,(V(\mathbf{x}_i)). \tag{7.5}$$

From Equation (7.1), we have

$$\mathrm{Var}\,(\widehat{V}(\mathbf{x}_i)) = \mathrm{Var}\left(\sum_{j=1}^{k} w_{ji} V(\mathbf{z}_j)\right) = \sum_{j=1}^{k} \sum_{l=1}^{k} w_{ji} w_{li} C_{jl}, \tag{7.6}$$

where C_{jl} is the covariance between $V(\mathbf{z}_j)$ and $V(\mathbf{z}_l)$. Similarly, we have

$$\mathrm{Cov}\,(\widehat{V}(\mathbf{x}_i), V(\mathbf{x}_i)) = \mathrm{Cov}\left(\sum_{j=1}^{k} w_{ji} V(\mathbf{z}_j), V(\mathbf{x}_i)\right)$$

$$= \sum_{j=1}^{k} w_{ji}\,\mathrm{Cov}\,(V(\mathbf{z}_j), V(\mathbf{x}_i))$$

$$= \sum_{j=1}^{k} w_{ji} D_{ji}, \tag{7.7}$$

where D_{ji} is the covariance between $V(\mathbf{z}_j)$ and $V(\mathbf{x}_i)$. We also assume that the random variables have the same variance, i.e.,

$$\text{Var}\,(V(\mathbf{x}_i)) = \sigma^2, \quad i = 1, 2, \ldots, n.$$

Hence we can write Equation (7.5) as

$$\text{Var}\,(R(\mathbf{x}_i)) = \sigma^2 + \sum_{j=1}^{k}\sum_{l=1}^{k} w_{ji}w_{li}C_{jl} - 2\sum_{j=1}^{k} w_{ji}D_{ji}. \qquad (7.8)$$

To minimize the variance of the estimation error subject to the constraint given in Equation (7.4), we use the method of Lagrange multipliers. That is, we minimize the following objective function:

$$f(W, \theta) = \sigma^2 + \sum_{j=1}^{k}\sum_{l=1}^{k} w_{ji}w_{li}C_{jl} - 2\sum_{j=1}^{k} w_{ji}D_{ji} + 2\theta\left(\sum_{j=1}^{k} w_{ji} - 1\right), \qquad (7.9)$$

where θ is the reference parameter and $W = (w_{li})$ is a $k \times n$ matrix containing the weights. By the method of Lagrange multipliers, we have

$$\frac{\partial f(W, \theta)}{w_{ji}} = 2w_{ji}\left(\sum_{l=1}^{k} w_{li}C_{jl} - D_{ji} + \theta\right) = 0, \qquad (7.10a)$$

$$\frac{\partial f(W, \theta)}{\theta} = 2\left(\sum_{j=1}^{k} w_{ji} - 1\right) = 0. \qquad (7.10b)$$

Equation (7.10) can be written in the following matrix form:

$$\begin{pmatrix} C_{11} & C_{12} & \cdots & C_{1k} & 1 \\ C_{21} & C_{22} & \cdots & C_{2k} & 1 \\ \vdots & \vdots & \vdots & \ddots & \vdots \\ C_{k1} & C_{k2} & \cdots & C_{kk} & 1 \\ 1 & 1 & \cdots & 1 & 0 \end{pmatrix} \begin{pmatrix} w_{1i} \\ w_{2i} \\ \vdots \\ w_{ki} \\ \theta \end{pmatrix} = \begin{pmatrix} D_{1i} \\ D_{2i} \\ \vdots \\ D_{ki} \\ 1 \end{pmatrix} \qquad (7.11)$$

for $i = 1, 2, \ldots, n$. The weights can be solved numerically from Equation (7.11).

By combining Equation (7.8) and Equation (7.11), we get the following variance of the estimation error:

$$\text{Var}\,(R(\mathbf{x}_i)) = \sigma^2 - \sum_{j=1}^{k} w_{ji}D_{ji} - \theta. \qquad (7.12)$$

To solve Equation (7.11), we need to model the covariances C_{jl} and D_{ji}.

This can be done by modeling the so-called variogram:

$$
\begin{aligned}
\beta_{jl} &= \frac{1}{2}E\left[(V(\mathbf{z}_j) - V(\mathbf{z}_l))^2\right] \\
&= \frac{1}{2}E\left[V(\mathbf{z}_j)^2\right] + \frac{1}{2}E\left[V(\mathbf{z}_l)^2\right] - E\left[V(\mathbf{z}_j)V(\mathbf{z}_l)\right] \\
&= \sigma^2 - C_{jl}.
\end{aligned}
\tag{7.13}
$$

The variogram describes the degree of spatial dependence of the random function.

Then Equation (7.11) can be rewritten as

$$
\begin{pmatrix}
\beta_{11} & \beta_{12} & \cdots & \beta_{1k} & 1 \\
\beta_{21} & \beta_{22} & \cdots & \beta_{2k} & 1 \\
\vdots & \vdots & \vdots & \ddots & \vdots \\
\beta_{k1} & \beta_{k2} & \cdots & \beta_{kk} & 1 \\
1 & 1 & \cdots & 1 & 0
\end{pmatrix}
\begin{pmatrix}
w_{1i} \\
w_{2i} \\
\vdots \\
w_{ki} \\
\theta
\end{pmatrix}
=
\begin{pmatrix}
\eta_{1i} \\
\eta_{2i} \\
\vdots \\
\eta_{ki} \\
1
\end{pmatrix}
\tag{7.14}
$$

for $i = 1, 2, \ldots, n$, where $\eta_{ji} = \sigma^2 - D_{ji}$. The variograms β_{jl} and η_{ji} can be modeled by a theoretical function as follows:

$$
\beta_{jl} = \gamma(\|\mathbf{z}_j - \mathbf{z}_l\|),
\tag{7.15a}
$$

$$
\eta_{ji} = \gamma(\|\mathbf{z}_j - \mathbf{x}_i\|),
\tag{7.15b}
$$

where $\gamma(\cdot)$ is a variogram model and $\|\cdot\|$ is a distance function. Thus, we see that a variogram can be viewed as a function of the distance between locations.

There are several commonly used variogram models. For example, the exponential variogram model is defined as

$$
\gamma(h) =
\begin{cases}
0, & \text{if } h = 0, \\
b + c\left(1 - \exp\left(-\frac{3h}{a}\right)\right), & \text{if } h > 0,
\end{cases}
\tag{7.16}
$$

where

- b is called the nugget effect, which provides a discontinuity at the origin,

- $b + c$ is called the sill, which is the variogram value for large distances, and

- a is called the range, which is a distance beyond which the variogram values become constant.

The spherical variogram model is defined as

$$
\gamma(h) =
\begin{cases}
0, & \text{if } h = 0, \\
b + c\left(1.5\frac{h}{a} - 0.5\left(\frac{h}{a}\right)^3\right), & \text{if } 0 < h \le a, \\
b + c, & \text{if } h > a.
\end{cases}
\tag{7.17}
$$

The Gaussian variogram model is defined as

$$\gamma(h) = \begin{cases} 0, & \text{if } h = 0, \\ b + c\left(1 - \exp\left(-\dfrac{3h^2}{a^2}\right)\right), & \text{if } h > 0. \end{cases} \tag{7.18}$$

The parameters a, b, and c of the variogram models can be estimated from the data. To estimate the parameters of a variogram model, we minimize the objective function

$$P(a, b, c) = \sum_{l=1}^{L} [\widehat{\gamma}(h_l) - \gamma(h_l; a, b, c)]^2, \tag{7.19}$$

where L is the number of bins of the distances. In the above equation, $\widehat{\gamma}(h_l)$ is the empirical variogram function and is defined as

$$\widehat{\gamma}(h) = \frac{1}{2|N(h)|} \sum_{(\mathbf{x}, \mathbf{y}) \in N(h)} [V(\mathbf{x}) - V(\mathbf{y})]^2, \tag{7.20}$$

where

$$N(h) = \{(\mathbf{x}, \mathbf{y}) : \|\mathbf{x} - \mathbf{y}\| = h, \mathbf{x} \in Z, \mathbf{y} \in Z\},$$

$|N(h)|$ is the number of elements in $N(h)$, and $Z = \{\mathbf{z}_1, \ldots, \mathbf{z}_k\}$.

When $b = 0$, the scale of c will not affect the weights W. However, it will affect the variance estimation given in Equation (7.12). Isaaks and Srivastava (1990) suggest that a good value for a is the 95th percentile of the distances between the locations $\mathbf{z}_1, \ldots, \mathbf{z}_k$.

7.2 Implementation

In this section, we implement the ordinary kriging method as described in the previous section. In particular, we implement two main functions: one for the ordinary kriging method and one for fitting variogram models.

First, let us implement the function for the ordinary kriging method. This function is created as follows:

```
1  okrig <- function(Z, y, X, varmodel) {
2    # Perform ordinary kriging prediction
3    #
4    # Args:
5    #   Z: a kxd matrix
6    #   y: a vector of length k
7    #   X: a nxd matrix
8    #   varmodel: a variogram model
```

```
 9    #
10    # Returns:
11    #   a vector of predicted values for X
12
13    k <- nrow(Z)
14    n <- nrow(X)
15    d <- ncol(Z)
16
17    # calculate distance matrix for Z
18    hZ <- matrix(0, nrow=k, ncol=k)
19    for(i in 1:k) {
20      hZ[i,] <- (apply((Z - matrix(Z[i,], nrow=k, ncol=d,
          byrow=T))^2, 1, sum))^0.5
21    }
22
23    # calculate distance matrix between Z and X
24    hD <- matrix(0, nrow=k, ncol=n)
25    for(i in 1:k) {
26      hD[i,] <- (apply((X - matrix(Z[i,], nrow=n, ncol=d,
          byrow=T))^2, 1, sum))^0.5
27    }
28
29    # construct kriging equation system
30    V <- matrix(1, nrow=k+1, ncol=k+1)
31    V[k+1, k+1] <- 0
32    V[1:k, 1:k] <- varmodel(hZ)
33
34    D <- matrix(1, nrow=k+1, ncol=n)
35    D[1:k,] <- varmodel(hD)
36
37    # solve equation
38    mW <- solve(V, D)
39
40    # perform prediction
41    mY <- matrix(0, nrow=k+1, ncol=1)
42    mY[1:k,1] <- y
43    yhat <- t(mW) %*% mY
44
45    return(yhat)
46 }
```

The name of the ordinary kriging function is okrig. The function has four
arguments as explained in the comments. In the above implementation, we
assume that the policies are represented by numerical vectors. Categorical
variables are converted to binary dummy variables and numerical variables
are normalized appropriately. Vectorization is used to speed up the distance
calculations.

The second main function is created as follows:

```
1 fitVarModel <- function(Z, y, vm, method) {
2   # fit a variogram model to data
3   #
4   # args:
5   #  Z: a kxd design matrix
6   #  y: a vector of length k
7   #  vm: a variogram function
8   #  method: fitting method ("default" or an integer
          specifying bins)
9   #
10  # returns:
11  #   c(a,b,c) parameters of a variogram model
12
13  # calculate distance matrix for Z
14  k <- nrow(Z)
15  d <- ncol(Z)
16  hZ <- matrix(0, nrow=k, ncol=k)
17  for(i in 1:k) {
18    hZ[i,] <- (apply((Z - matrix(Z[i,], nrow=k, ncol=d,
          byrow=T))^2, 1, sum))^0.5
19  }
20  vD <- hZ[upper.tri(hZ)]
21  da <- quantile(vD, 0.95)
22  db <- 0
23  dc <- var(y)
24
25  if(method == "default") {
26    return(c(da, db, dc))
27  }
28
29  nBin <- method
30  if(nBin <=3) {
31    stop("number of bins <=3")
32  }
33
34  dMin <- min(vD)
35  dMax <- max(vD) + 1
36  dBandWidth <- (dMax - dMin) / nBin
37  vh <- c()
38  vy <- c()
39  for(j in 1:nBin) {
40    dL <- dMin + (j-1)* dBandWidth
41    dU <- dL + dBandWidth
42    ind <- which(vD >= dL & vD < dU)
43    if(length(ind) > 0) {
44      dSum <- 0
45      for(t in ind) {
46        cInd <- ceiling(sqrt(2*t + 0.25)+0.5)
```

```
47            rInd <- t - (cInd-1)*(cInd-2)/2
48            dSum <- dSum + (y[rInd] - y[cInd])^2
49          }
50          vh <- c(vh, (dL + dU)/2)
51          vy <- c(vy, dSum / (2 * length(ind)))
52      }
53  }
54
55    plot(vh,vy,main="Fit variogram model", xlab="h", ylab=
          "gamma(h)")
56    fit <- nls(vy~vm(vh,a,b,c),start=list(a=da,b=db,c=dc))
57
58    res <- coef(fit)
59    # plot empirical and fitted variogram models
60    curve(vm(x,res[1], res[2], res[3]),add=TRUE)
61
62    return(res)
63  }
```

The name of the function is `fitVarModel`. It can be used to fit a variogram model to a given dataset. This function also has four arguments, which are explained in the comments. The R function `nls` is used to fit a variogram model to the empirical variogram.

The following R functions implement three theoretical variogram models:

```
1  expVM <- function(h, a, b, c) {
2    res <- h
3    ind <- h>0
4    res[!ind] <- 0
5    res[ind] <- b + c*(1-exp(-3*h[ind]/a))
6    return(res)
7  }
8
9  sphVM <- function(h, a, b, c) {
10    res <- h
11    ind <- h==0
12    res[ind] <- 0
13    ind <- h>0 & h<= a
14    res[ind] <- b + c*(1.5*h[ind]/a-0.5*(h[ind]/a)^3)
15    ind <- h > a
16    res[ind] <- b + c
17    return(res)
18  }
19
20  gauVM <- function(h, a, b, c) {
21    res <- h
22    ind <- h>0
23    res[!ind] <- 0
24    res[ind] <- b + c*(1-exp(-3*(h[ind]/a)^2))
```

```
25    return(res)
26 }
```

As we can see from the above code, vectorization is used in the implementation of all three functions. The output has the same shape as the input h.

7.3   Applications

In this section, we apply the ordinary kriging method to predict the fair market values of the synthetic policies described in Appendix A. To assess the quality of the predictions, we use the following two validation measures: the percentage error at the portfolio level and the R-squared.

For $i = 1, 2, \ldots, n$, let y_i be the fair market value of the ith variable annuity policy that is calculated by Monte Carlo simulation and let $\widehat{y}_i$ be the fair market value predicted by the ordinary kriging method. Then the percentage error at the portfolio level is defined as

$$PE = \frac{\sum_{i=1}^{n}(\widehat{y}_i - y_i)}{\sum_{i=1}^{n} y_i}. \tag{7.21}$$

The R-squared is defined as

$$R^2 = 1 - \frac{\sum_{i=1}^{n}(\widehat{y}_i - y_i)^2}{\sum_{i=1}^{n}(y_i - \bar{y})^2}, \tag{7.22}$$

where $\bar{y}$ is the average fair market value, i.e.,

$$\bar{y} = \frac{1}{n}\sum_{i=1}^{n} y_i.$$

The percentage error measures the aggregate accuracy of the result because the errors at the individual policy level can offset each other. The lower the absolute value of PE, the better the result. The R^2 measures the accuracy of the result at the individual policy level. The higher the R^2, the better the result.

Now let us load the inforce and the Greeks into R by running the following code:

```
1 inforce <- read.csv("inforce.csv")
2
3 vNames <- c("gbAmt", "gmwbBalance", "withdrawal", paste
     ("FundValue", 1:10, sep=""))
4
5 age <- with(inforce, (currentDate-birthDate)/365)
```

```
 6 ttm <- with(inforce, (matDate - currentDate)/365)
 7
 8 datN <- cbind(inforce[,vNames], data.frame(age=age, ttm=
     ttm))
 9 datC <- inforce[,c("gender", "productType")]
10 dat <- cbind(datN, datC)
11
12 greek <- read.csv("Greek.csv")
13 greek <- greek[order(greek$recordID),]
```

In the above code, we assume that the inforce file and the Greek file are saved in the working directory.

Once the policies are loaded, we convert the categorical variables to numerical as follows:

```
 1 > X <- model.matrix( ~ ., data=dat)[,-1]
 2 > colnames(X)
 3  [1] "gbAmt"            "gmwbBalance"      "withdrawal"
 4  [4] "FundValue1"       "FundValue2"       "FundValue3"
 5  [7] "FundValue4"       "FundValue5"       "FundValue6"
 6 [10] "FundValue7"       "FundValue8"       "FundValue9"
 7 [13] "FundValue10"      "age"              "ttm"
 8 [16] "genderM"          "productTypeABRU"  "
     productTypeABSU"
 9 [19] "productTypeDBAB" "productTypeDBIB"  "
     productTypeDBMB"
10 [22] "productTypeDBRP" "productTypeDBRU"  "
     productTypeDBSU"
11 [25] "productTypeDBWB" "productTypeIBRP"  "
     productTypeIBRU"
12 [28] "productTypeIBSU" "productTypeMBRP"  "
     productTypeMBRU"
13 [31] "productTypeMBSU" "productTypeWBRP"  "
     productTypeWBRU"
14 [34] "productTypeWBSU"
15 > summary(X[,1:10])
16     gbAmt            gmwbBalance          withdrawal
17  Min.   : 50002   Min.   :      0   Min.   :      0
18  1st Qu.:179759   1st Qu.:      0   1st Qu.:      0
19  Median :303525   Median :      0   Median :      0
20  Mean   :313507   Mean   : 36141   Mean   : 21928
21  3rd Qu.:427544   3rd Qu.:      0   3rd Qu.:      0
22  Max.   :989205   Max.   :499709   Max.   :499586
23    FundValue1         FundValue2          FundValue3
24  Min.   :     0   Min.   :     0   Min.   :     0
25  1st Qu.:     0   1st Qu.:     0   1st Qu.:     0
26  Median :  8299   Median :  8394   Median :  4942
27  Mean   : 26611   Mean   : 26045   Mean   : 17391
28  3rd Qu.: 39209   3rd Qu.: 38463   3rd Qu.: 24251
```

```
29  Max.    :921549   Max.    :844323   Max.    :580753
30    FundValue4         FundValue5         FundValue6
31  Min.   :       0   Min.   :       0   Min.   :       0
32  1st Qu.:       0   1st Qu.:       0   1st Qu.:       0
33  Median :    4225   Median :    7248   Median :    8556
34  Mean   :   14507   Mean   :   21041   Mean   :   26570
35  3rd Qu.:   20756   3rd Qu.:   32112   3rd Qu.:   39241
36  Max.   :  483937   Max.   :  494382   Max.   :  872707
37    FundValue7
38  Min.   :       0
39  1st Qu.:       0
40  Median :    6602
41  Mean   :   21506
42  3rd Qu.:   31088
43  Max.   :  634819
```

Some variables have different scales and magnitudes. Variables with large magnitudes can dominate the distance. To avoid variables dominating the distance, we normalize all the variables as follows:

```
1 > vMin <- apply(X, 2, min)
2 > vMax <- apply(X, 2, max)
3 > X <- (X - matrix(vMin, nrow=nrow(X), ncol= ncol(X),
      byrow=TRUE)) / matrix(vMax-vMin, nrow=nrow(X), ncol=
      ncol(X), byrow=TRUE)
4 > summary(X[,1:10])
5      gbAmt            gmwbBalance          withdrawal
6  Min.   :0.0000   Min.    :0.00000   Min.    :0.00000
7  1st Qu.:0.1382   1st Qu.:0.00000   1st Qu.:0.00000
8  Median :0.2699   Median :0.00000   Median :0.00000
9  Mean   :0.2806   Mean   :0.07232   Mean   :0.04389
10 3rd Qu.:0.4020   3rd Qu.:0.00000   3rd Qu.:0.00000
11 Max.   :1.0000   Max.    :1.00000   Max.    :1.00000
12    FundValue1         FundValue2         FundValue3
13 Min.   :0.000000   Min.    :0.000000   Min.    :0.000000
14 1st Qu.:0.000000   1st Qu.:0.000000   1st Qu.:0.000000
15 Median :0.009006   Median :0.009942   Median :0.008509
16 Mean   :0.028877   Mean   :0.030847   Mean   :0.029946
17 3rd Qu.:0.042547   3rd Qu.:0.045555   3rd Qu.:0.041759
18 Max.   :1.000000   Max.    :1.000000   Max.    :1.000000
19    FundValue4         FundValue5         FundValue6
20 Min.   :0.000000   Min.    :0.00000   Min.    :0.000000
21 1st Qu.:0.000000   1st Qu.:0.00000   1st Qu.:0.000000
22 Median :0.008731   Median :0.01466   Median :0.009804
23 Mean   :0.029978   Mean   :0.04256   Mean   :0.030445
24 3rd Qu.:0.042890   3rd Qu.:0.06495   3rd Qu.:0.044965
25 Max.   :1.000000   Max.    :1.00000   Max.    :1.000000
26    FundValue7
27 Min.   :0.00000
```

```
28   1st Qu.:0.00000
29   Median :0.01040
30   Mean   :0.03388
31   3rd Qu.:0.04897
32   Max.   :1.00000
```

7.3.1   Ordinary Kriging with Latin Hypercube Sampling

We are ready to use the ordinary kriging method to make predictions. First, we use the ordinary kriging method based on the representative policies selected by Latin hypercube sampling described in Chapter 4. We read the indices of the representative policies and fit the exponential variogram model to the data as follows:

```
1 > S <- read.table("lhs.csv", sep=",")
2 > S <- S[,2]
3 > Z <- X[S,]
4 > y <- greek$fmv[S]/1000
5 >
6 > {t1 <- proc.time()
7 + res <- fitVarModel(Z, y, expVM, 100)
8 + proc.time()-t1}
9    user   system elapsed
10   0.39     0.04    0.44
```

In the above code, we used the R function proc.time to measure the runtime of the variogram fitting. From the output, we see that fitting the variogram is quite fast. The empirical variogram and the fitted theoretical variogram model are shown in Figure 7.1.

We use the ordinary kriging method implemented in the previous section as follows:

```
1 > {t1 <- proc.time()
2 + yhat <- okrig(Z, y, X, function(h) {(expVM(h, res[1],
      res[2], res[3]))})
3 + proc.time()-t1}
4    user   system elapsed
5   217.78   13.82   242.47
6 > calMeasure(greek$fmv/1000, yhat)
7 $pe
8 [1] -0.3761338
9
10 $r2
11 [1] 0.3824283
```

From the output, we see that it took the function about 4 minutes to finish due to the large number of distance calculations. The percentage error is about

Fit variogram model

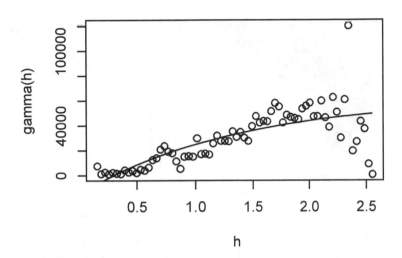

FIGURE 7.1: The empirical variogram and the fitted exponential variogram model. The representative policies are selected by Latin hypercube sampling.

-37%, which is relatively not good. The R^2 is also low. These results indicate that the representative policies selected by Latin hypercube sampling are not good.

To visualize the goodness-of-fit of the ordinary kriging model, we can plot the fair market values as follows:

```
1 > png("expqq.png", width=8, height=4, units="in", res
    =300)
2 > par(mfrow=c(1,2), mar=c(4,4,1,1))
3 > plot(greek$fmv/1000, yhat, xlab="FMV(MC)", ylab="FMV(
    OK)")
4 > abline(0,1)
5 > qqplot(greek$fmv/1000, yhat, xlab="FMV(MC)", ylab="FMV
    (OK)")
6 > abline(0,1)
7 > dev.off()
8 null device
9           1
```

In the above code, we save the resulting plot to a PNG figure, which has a smaller size than a PDF file. The resulting plots are shown in Figure 7.2. From the plots, we see that the model does not fit the left tail well. The scatter plot shows that the ordinary kriging method predicted many large negative values; while not many values calculated by Monte Carlo are negative.

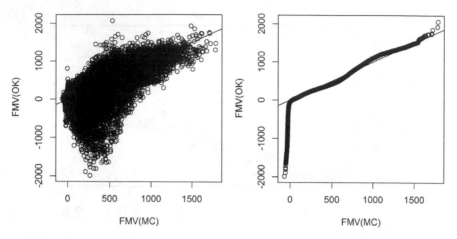

FIGURE 7.2: A scatter plot and a QQ plot of the fair market values calculated by Monte Carlo simulation and those predicted by ordinary kriging with an exponential variogram. The representative policies are selected by Latin hypercube sampling.

To see how the ordinary kriging method performs based on the spherical variogram model, we proceed as follows:

```
 1 > {t1 <- proc.time()
 2 + res <- fitVarModel(Z, y, sphVM, 100)
 3 + proc.time()-t1}
 4    user   system elapsed
 5    0.39    0.04    0.46
 6 > res
 7              a            b            c
 8      2.238043  -12424.305722   58737.822929
 9 >
10 > {t1 <- proc.time()
11 + yhat <- okrig(Z, y, X, function(h) {(sphVM(h, res[1],
        res[2], res[3]))})
12 + proc.time()-t1}
13    user   system elapsed
14   253.19   15.69   272.61
15 > calMeasure(greek$fmv/1000, yhat)
16 $pe
17 [1]  -0.4599771
18
19 $r2
20 [1]  -0.2291823
21
```

```
22 >
23 > png("sphqq.png", width=8, height=4, units="in", res
      =300)
24 > par(mfrow=c(1,2), mar=c(4,4,1,1))
25 > plot(greek$fmv/1000, yhat, xlab="FMV(MC)", ylab="FMV(
      OK)")
26 > abline(0,1)
27 > qqplot(greek$fmv/1000, yhat, xlab="FMV(MC)", ylab="FMV
      (OK)")
28 > abline(0,1)
29 > dev.off()
30 windows
31       2
```

The above output shows that ordinary kriging with the spherical variogram
model produced even worse results than when the exponential variogram
model was used. For example, the R^2 based on the spherical variogram model
is negative. The scatter plot and the QQ plot are shown in Figure 7.3. From
the figure, we see that when the spherical variogram model is used, ordinary
kriging does not fit the data well.

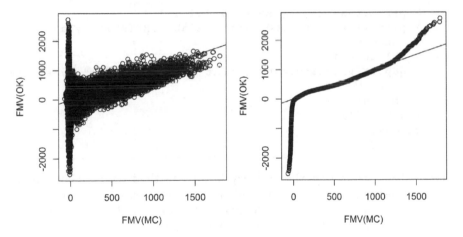

FIGURE 7.3: A scatter plot and a QQ plot of the fair market values calculated by Monte Carlo simulation and those predicted by ordinary kriging
with a spherical variogram. The representative policies are selected by Latin
hypercube sampling.

The following output shows the performance of the ordinary kriging
method with a Gaussian variogram:

```
1 > {t1 <- proc.time()
2 + res <- fitVarModel(Z, y, gauVM, 100)
3 + proc.time()-t1}
```

```
 4      user   system elapsed
 5      0.39     0.02    0.42
 6 > res
 7             a           b           c
 8      2.039751 -2563.415448 50597.212708
 9 >
10 > {t1 <- proc.time()
11 + yhat <- okrig(Z, y, X, function(h) {(gauVM(h, res[1],
       res[2], res[3]))})
12 + proc.time()-t1}
13      user   system elapsed
14   252.72    14.71  272.99
15 > calMeasure(greek$fmv/1000, yhat)
16 $pe
17 [1] -1.038825
18
19 $r2
20 [1] -2.256726
21
22 >
23 > png("gauqq.png", width=8, height=4, units="in", res
       =300)
24 > par(mfrow=c(1,2), mar=c(4,4,1,1))
25 > plot(greek$fmv/1000, yhat, xlab="FMV(MC)", ylab="FMV(
       OK)")
26 > abline(0,1)
27 > qqplot(greek$fmv/1000, yhat, xlab="FMV(MC)", ylab="FMV
       (OK)")
28 > abline(0,1)
29 > dev.off()
30 windows
31         2
```

The output shows that the ordinary kriging method with a Gaussian variogram performs the worst. Figure 7.4 shows the scatter plot and the QQ plot of the fair market values. The figure shows that ordinary kriging with a Gaussian variogram does not fit the left tail well.

7.3.2 Ordinary Kriging with Conditional Latin Hypercube Sampling

To test the performance of ordinary kriging with representative policies selected by conditional Latin hypercube sampling, we need to change the indices of the representative policies. This can be done as follows:

```
1 > S <- read.table("clhs.csv", sep=",")
2 > S <- S[,2]
3 > Z <- X[S,]
```

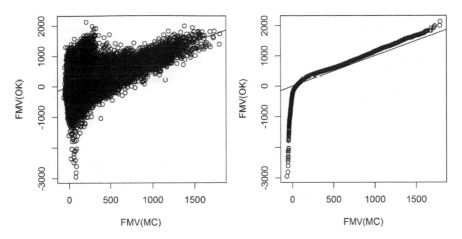

FIGURE 7.4: A scatter plot and a QQ plot of the fair market values calculated by Monte Carlo simulation and those predicted by ordinary kriging
with a Gaussian variogram. The representative policies are selected by Latin
hypercube sampling.

```
 4 > y <- greek$fmv[S]/1000
 5 >
 6 > {t1 <- proc.time()
 7 + res <- fitVarModel(Z, y, expVM, 100)
 8 + proc.time()-t1}
 9    user   system  elapsed
10    0.54     0.47     1.03
11 >
12 > {t1 <- proc.time()
13 + yhat <- okrig(Z, y, X, function(h) {(expVM(h, res[1],
      res[2], res[3]))})
14 + proc.time()-t1}
15    user   system  elapsed
16  242.49    16.59   292.80
17 > calMeasure(greek$fmv/1000, yhat)
18 $pe
19 [1]  -0.03069026
20
21 $r2
22 [1]  0.8069264
23
24 >
25 > png("expclhs.png", width=8, height=4, units="in", res
      =300)
26 > dev.new(width=8, height=4)
27 > par(mfrow=c(1,2), mar=c(4,4,1,1))
```

```
28 > plot(greek$fmv/1000, yhat, xlab="FMV(MC)", ylab="FMV(
      OK)")
29 > abline(0,1)
30 > qqplot(greek$fmv/1000, yhat, xlab="FMV(MC)", ylab="FMV
      (OK)")
31 > abline(0,1)
32 > dev.off()
33 windows
34       2
```

The validation measures in the output show that ordinary kriging with conditional Latin hypercube sampling produced much better results than that with Latin hypercube sampling. However, the scatter plot and the QQ plot given in Figure 7.5 show that the tails are not fitted well by the model.

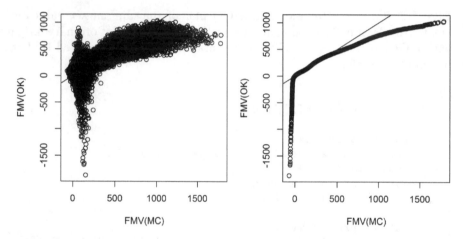

FIGURE 7.5: A scatter plot and a QQ plot of the fair market values calculated by Monte Carlo simulation and those predicted by ordinary kriging with an exponential variogram. The representative policies are selected by conditional Latin hypercube sampling.

To see the performance of ordinary kriging with a spherical variogram, we change the variogram model as follows:

```
1 > {t1 <- proc.time()
2 + res <- fitVarModel(Z, y, sphVM, 100)
3 + proc.time()-t1}
4     user   system elapsed
5     0.47     0.03    0.81
6 >
7 > {t1 <- proc.time()
8 + yhat <- okrig(Z, y, X, function(h) {(sphVM(h, res[1],
      res[2], res[3]))})
```

```
 9 + proc.time()-t1}
10    user   system elapsed
11  235.09    14.81   256.93
12 > calMeasure(greek$fmv/1000, yhat)
13 $pe
14 [1] -0.0259222
15
16 $r2
17 [1] 0.8828032
18
19 >
20 > png("sphclhs.png", width=8, height=4, units="in", res
     =300)
21 > par(mfrow=c(1,2), mar=c(4,4,1,1))
22 > plot(greek$fmv/1000, yhat, xlab="FMV(MC)", ylab="FMV(
     OK)")
23 > abline(0,1)
24 > qqplot(greek$fmv/1000, yhat, xlab="FMV(MC)", ylab="FMV
     (OK)")
25 > abline(0,1)
26 > dev.off()
27 null device
28            1
```

The above output shows that when the spherical variogram model and conditional Latin hypercube sampling were used, the ordinary kriging produced very good results. The R^2 is about 0.88 and the percentage error is around 2.6%. The good results are also confirmed by the scatter plot and the QQ plot shown in Figure 7.6.

The performance of ordinary kriging with a Gaussian variogram is shown below:

```
 1 > {t1 <- proc.time()
 2 + res <- fitVarModel(Z, y, gauVM, 100)
 3 + proc.time()-t1}
 4    user   system elapsed
 5    0.33    0.04    0.44
 6 >
 7 > {t1 <- proc.time()
 8 + yhat <- okrig(Z, y, X, function(h) {(gauVM(h, res[1],
     res[2], res[3]))})
 9 + proc.time()-t1}
10    user   system elapsed
11  245.78    14.20   270.97
12 > calMeasure(greek$fmv/1000, yhat)
13 $pe
14 [1] -0.0238202
15
16 $r2
```

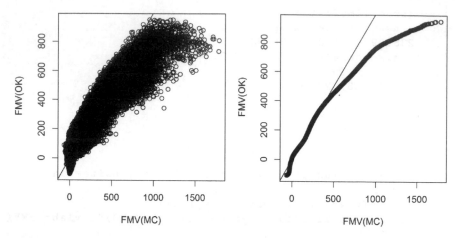

FIGURE 7.6: A scatter plot and a QQ plot of the fair market values calculated by Monte Carlo simulation and those predicted by ordinary kriging with a spherical variogram. The representative policies are selected by conditional Latin hypercube sampling.

```
17 [1] 0.8872577
18
19 >
20 > png("gauclhs.png", width=8, height=4, units="in", res
      =300)
21 > par(mfrow=c(1,2), mar=c(4,4,1,1))
22 > plot(greek$fmv/1000, yhat, xlab="FMV(MC)", ylab="FMV(
      OK)")
23 > abline(0,1)
24 > qqplot(greek$fmv/1000, yhat, xlab="FMV(MC)", ylab="FMV
      (OK)")
25 > abline(0,1)
26 > dev.off()
27 null device
28            1
```

The validation measures show that the results based on a Gaussian variogram are a little bit better than those based on a spherical variogram. The scatter plot and the QQ plot shown in Figure 7.7 are similar to those shown in Figure 7.6.

7.3.3   Ordinary Kriging with Hierarchical k-Means

In this subsection, we test the performance of ordinary kriging with representative policies selected by the hierarchical k-means algorithm. We use the indices of the representative policies produced in Chapter 6.

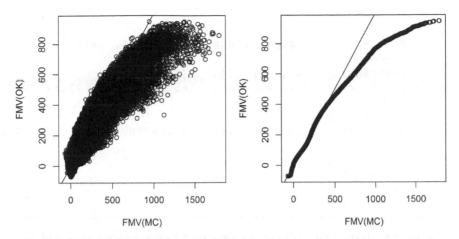

FIGURE 7.7: A scatter plot and a QQ plot of the fair market values calculated by Monte Carlo simulation and those predicted by ordinary kriging with a Gaussian variogram. The representative policies are selected by conditional Latin hypercube sampling.

The following R output shows the performance of ordinary kriging with an exponential variogram and the representative policies selected by hierarchical k-means:

```
 1 > S <- read.table("hkmeans.csv", sep=",")
 2 > S <- S[,2]
 3 > Z <- X[S,]
 4 > y <- greek$fmv[S]/1000
 5 >
 6 > {t1 <- proc.time()
 7 + res <- fitVarModel(Z, y, expVM, 100)
 8 + proc.time()-t1}
 9    user  system elapsed
10    0.36    0.03    0.43
11 >
12 > {t1 <- proc.time()
13 + yhat <- okrig(Z, y, X, function(h) {(expVM(h, res[1],
        res[2], res[3]))})
14 + proc.time()-t1}
15    user  system elapsed
16  221.91   13.56  237.65
17 > calMeasure(greek$fmv/1000, yhat)
18 $pe
19 [1] 0.02525438
20
21 $r2
22 [1] 0.9217258
```

```
23
24 >
25 > png("exphkmean.png", width=8, height=4, units="in",
     res=300)
26 > par(mfrow=c(1,2), mar=c(4,4,1,1))
27 > plot(greek$fmv/1000, yhat, xlab="FMV(MC)", ylab="FMV(
     OK)")
28 > abline(0,1)
29 > qqplot(greek$fmv/1000, yhat, xlab="FMV(MC)", ylab="FMV
     (OK)")
30 > abline(0,1)
31 > dev.off()
32 null device
33                 1
```

The output shows that ordinary kriging based on representative policies selected by hierarchical k-means produced better results than those based on other experimental design methods. The R^2 is around 0.92. The scatter and QQ plots shown in Figure 7.8 indicate that the model has a good overall fit. However, the model does not fit the right tail well.

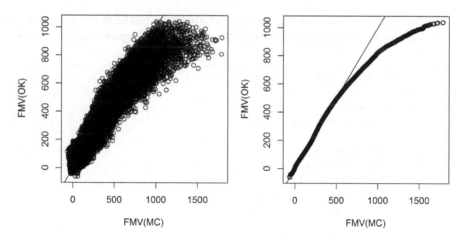

FIGURE 7.8: A scatter plot and a QQ plot of the fair market values calculated by Monte Carlo simulation and those predicted by ordinary kriging with an exponential variogram. The representative policies are selected by hierarchical k-means.

The following output shows the performance of ordinary kriging when the spherical variogram model is used:

```
1 > {t1 <- proc.time()
2 + res <- fitVarModel(Z, y, sphVM, 100)
3 + proc.time()-t1}
```

```
 4      user   system  elapsed
 5      0.52     0.39     0.92
 6 >
 7 > {t1 <- proc.time()
 8 + yhat <- okrig(Z, y, X, function(h) {(sphVM(h, res[1],
        res[2], res[3]))})
 9 + proc.time()-t1}
10      user   system  elapsed
11   231.93    13.73   248.00
12 > calMeasure(greek$fmv/1000, yhat)
13 $pe
14 [1]  0.02664327
15
16 $r2
17 [1]  0.9207789
18
19 >
20 > png("sphhkmean.png", width=8, height=4, units="in",
        res=300)
21 > par(mfrow=c(1,2), mar=c(4,4,1,1))
22 > plot(greek$fmv/1000, yhat, xlab="FMV(MC)", ylab="FMV(
        OK)")
23 > abline(0,1)
24 > qqplot(greek$fmv/1000, yhat, xlab="FMV(MC)", ylab="FMV
        (OK)")
25 > abline(0,1)
26 > dev.off()
27 windows
28        2
```

We obtained similar results as before. The scatter and QQ plots are shown in Figure 7.9.

We also would like to see the performance of ordinary kriging with a Gaussian variogram. We change the variogram model as follows:

```
 1 > {t1 <- proc.time()
 2 + res <- fitVarModel(Z, y, gauVM, 100)
 3 + proc.time()-t1}
 4      user   system  elapsed
 5      0.37     0.00     0.38
 6 >
 7 > {t1 <- proc.time()
 8 + yhat <- okrig(Z, y, X, function(h) {(gauVM(h, res[1],
        res[2], res[3]))})
 9 + proc.time()-t1}
10      user   system  elapsed
11   234.71    15.30   254.65
12 > calMeasure(greek$fmv/1000, yhat)
13 $pe
```

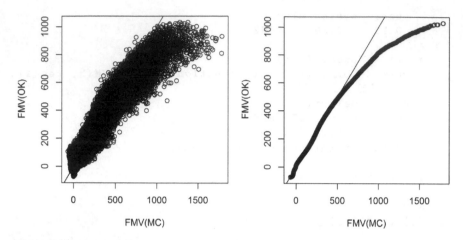

FIGURE 7.9: A scatter plot and a QQ plot of the fair market values calculated by Monte Carlo simulation and those predicted by ordinary kriging with a spherical variogram. The representative policies are selected by hierarchical k-means.

```
14  [1]  0.02980923
15
16  $r2
17  [1]  0.916729
18
19  >
20  > png("gauhkmean.png", width=8, height=4, units="in",
        res=300)
21  > par(mfrow=c(1,2), mar=c(4,4,1,1))
22  > plot(greek$fmv/1000, yhat, xlab="FMV(MC)", ylab="FMV(
        OK)")
23  > abline(0,1)
24  > qqplot(greek$fmv/1000, yhat, xlab="FMV(MC)", ylab="FMV
        (OK)")
25  > abline(0,1)
26  > dev.off()
27  windows
28         2
```

The validation measures shown in the above output show that the results based on a Gaussian variogram are not as good as those based on exponential and spherical variograms. Figure 7.10 shows the scatter and the QQ plots of this case. Again, the model does not fit the right tail well.

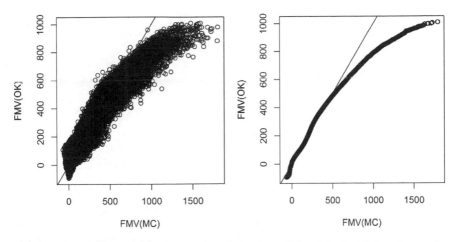

FIGURE 7.10: A scatter plot and a QQ plot of the fair market values calculated by Monte Carlo simulation and those predicted by ordinary kriging with a Gaussian variogram. The representative policies are selected by hierarchical k-means.

TABLE 7.1: Validation measures of ordinary kriging with different combinations of variogram models and experimental design methods.

	PE			R^2		
	LHS	cLHS	Hkmeans	LHS	cLHS	Hkmeans
Exponential	−37.61%	−3.07%	2.53%	0.3824	0.8069	0.9217
Spherical	−46.00%	−2.59%	2.66%	−0.2292	0.8828	0.9208
Gaussian	−103.88%	−2.38%	2.98%	−2.2567	0.8873	0.9167

7.4 Summary

In this chapter, we introduced the ordinary kriging method and three theoretical variogram models: exponential, spherical, and Gaussian. These variogram models are used to model the degree of spatial dependence between locations. We also tested the performance of the ordinary kriging method with various combinations of experimental design methods and variogram models. The results are summarized in Table 7.1. The results show that ordinary kriging produced the best results when the hierarchical k-means and the exponential variogram model were used.

FIGURE 4.10 [illegible] (left) and [illegible] of the [illegible] calculated by Monte Carlo simulation and those predicted by ordinary kriging with a Gaussian variogram. The representative particles are selected by [illegible] k-means.

TABLE 4.1 Validation [illegible] of [illegible] with different [illegible] of [illegible] models and experimental design methods.

	[illegible]	[illegible]	[illegible]	[illegible]	Hammersley	[illegible]	[illegible]
[illegible]	[illegible]	[illegible]	[illegible]	[illegible]	[illegible]	[illegible]	[illegible]
[illegible]	[illegible]	[illegible]	[illegible]	[illegible]	[illegible]	[illegible]	[illegible]
[illegible]	[illegible]	[illegible]	[illegible]	[illegible]	[illegible]	[illegible]	[illegible]

[illegible]

[illegible]

8

Universal Kriging

In Chapter 7, we introduced the ordinary kriging method, which is one of the commonly used kriging techniques. Ordinary kriging assumes a stationary random function model, where the mean is constant across different locations. In this chapter, we introduce universal kriging, which allows for a smoothly varying and nonstationary trend. Ordinary kriging can be viewed as a special case of universal kriging when the trend is constant.

8.1 Description of the Model

The universal kriging method is similar to the ordinary kriging method in that both methods use a random function to model the data. In the universal kriging method, the random function is divided into two components: a linear combination of deterministic functions and a random component (Wackernagel, 2003, Chapter 38).

Mathematically, the universal kriging method is described as follows. Let V be the random function. Denote $\mathbf{z}_1, \mathbf{z}_2, \ldots, \mathbf{z}_k$ as the locations where the values of the random function are known. Denote $\mathbf{x}_1, \mathbf{x}_2, \ldots, \mathbf{x}_n$ as the locations where the values of the random function are unknown. At an arbitrary location $\mathbf{x}$, the random function consists of a deterministic component $m(\mathbf{x})$ and a second-order stationary random function $Y(\mathbf{x})$:

$$V(\mathbf{x}) = m(\mathbf{x}) + Y(\mathbf{x}). \tag{8.1}$$

The first component $m(\mathbf{x})$ is also referred to as the drift. In addition, the drift is assumed to be a linear combination of deterministic functions:

$$m(\mathbf{x}) = \sum_{l=0}^{J} a_l f_l(\mathbf{x}), \tag{8.2}$$

where $a_0, a_1, \ldots, a_J$ are non-zero coefficients and $f_0(\mathbf{x}) = 1$. The functions $f_0, f_1, \ldots, f_J$ are called basis functions. The second component is usually assumed to have a mean of zero and a covariance function $C(h)$. As a result,

$$E[V(\mathbf{x})] = m(\mathbf{x}). \tag{8.3}$$

For $i = 1, 2, \ldots, n$, the random variable at the unknown location $\mathbf{x}_i$ is estimated as

$$\widehat{V}(\mathbf{x}_i) = \sum_{j=1}^{k} w_{ji} V(\mathbf{z}_j).\tag{8.4}$$

To ensure unbiased estimates, we impose the following condition:

$$E[V(\mathbf{x}_i) - \widehat{V}(\mathbf{x}_i)] = 0,$$

which gives

$$m(\mathbf{x}_i) - \sum_{j=1}^{k} w_{ji} m(\mathbf{z}_j) = 0.$$

Plugging Equation (8.2) into the above equation leads to

$$\sum_{l=0}^{J} a_l \left(f_l(\mathbf{x}_i) - \sum_{j=1}^{k} w_{ji} f_l(\mathbf{z}_j) \right) = 0.$$

Since a_l's are non-zero, we have

$$\sum_{j=1}^{k} w_{ji} f_l(\mathbf{z}_j) = f_l(\mathbf{x}_i), \quad l = 0, 1, \ldots, J.\tag{8.5}$$

The conditions on the weights given in Equation (8.5) are called universality conditions. Since $f_0(\mathbf{x}) = 0$, we have the usual condition:

$$\sum_{j=1}^{k} w_{ji} = 1.$$

Under the universality conditions, the estimation error becomes

$$R(\mathbf{x}_i) = \widehat{V}(\mathbf{x}_i) - V(\mathbf{x}_i) = \sum_{j=1}^{k} w_{ji} V(\mathbf{z}_j) - V(\mathbf{x}_i)$$

$$= \sum_{j=1}^{k} w_{ji} Y(\mathbf{z}_j) + \sum_{j=1}^{k} w_{ji} m(\mathbf{z}_j) - Y(\mathbf{x}_i) - m(\mathbf{x}_i)$$

$$= \sum_{j=1}^{k} w_{ji} Y(\mathbf{z}_j) - Y(\mathbf{x}_i).\tag{8.6}$$

The variance of the estimation error is calculated as follows:

$$\mathrm{Var}\,(R(\mathbf{x}_i)) = \mathrm{Var}\,(\widehat{Y}(\mathbf{x}_i)) - 2\,\mathrm{Cov}\,(\widehat{Y}(\mathbf{x}_i), Y(\mathbf{x}_i)) + \mathrm{Var}\,(Y(\mathbf{x}_i)).\tag{8.7}$$

From Equation (8.4), we have

$$\text{Var}\,(\widehat{Y}(\mathbf{x}_i)) = \text{Var}\left(\sum_{j=1}^{k} w_{ji}Y(\mathbf{z}_j)\right) = \sum_{j=1}^{k}\sum_{l=1}^{k} w_{ji}w_{li}C_{jl}, \qquad (8.8)$$

where C_{jl} is the covariance between $Y(\mathbf{z}_j)$ and $Y(\mathbf{z}_l)$. Similarly, we have

$$\text{Cov}\,(\widehat{Y}(\mathbf{x}_i), Y(\mathbf{x}_i)) = \text{Cov}\left(\sum_{j=1}^{k} w_{ji}Y(\mathbf{z}_j), Y(\mathbf{x}_i)\right)$$

$$= \sum_{j=1}^{k} w_{ji}\,\text{Cov}\,(Y(\mathbf{z}_j), Y(\mathbf{x}_i))$$

$$= \sum_{j=1}^{k} w_{ji}D_{ji}, \qquad (8.9)$$

where D_{ji} is the covariance between $Y(\mathbf{z}_j)$ and $Y(\mathbf{x}_i)$. We also assume that the random components have the same variance, i.e.,

$$\text{Var}\,(Y(\mathbf{x}_i)) = \sigma^2.$$

Hence we can write Equation (8.7) as

$$\text{Var}\,(R(\mathbf{x}_i)) = \sigma^2 + \sum_{j=1}^{k}\sum_{l=1}^{k} w_{ji}w_{li}C_{jl} - 2\sum_{j=1}^{k} w_{ji}D_{ji}. \qquad (8.10)$$

To minimize the variance of the estimation error subject to the constraint given in Equation (8.5), we use the method of Lagrange multipliers. That is, we minimize the following objective function:

$$f(W, \boldsymbol{\theta}) = \sigma^2 + \sum_{j=1}^{k}\sum_{l=1}^{k} w_{ji}w_{li}C_{jl} - 2\sum_{j=1}^{k} w_{ji}D_{ji} +$$

$$2\sum_{l=0}^{J} \theta_l \left(\sum_{j=1}^{k} w_{ji}f_l(\mathbf{z}_j) - f_l(\mathbf{x}_i)\right), \qquad (8.11)$$

where $\boldsymbol{\theta} = (\theta_0, \theta_1, \ldots, \theta_J)$ is a vector of reference parameters and W is a $k \times n$ matrix containing the weights. By the method of Lagrange multipliers, we have

$$\frac{\partial f(W, \boldsymbol{\theta})}{w_{ji}} = 2w_{ji}\left(\sum_{l=1}^{k} w_{li}C_{jl} - D_{ji} + \sum_{l=0}^{J} \theta_l f_l(\mathbf{z}_j)\right) = 0, \qquad (8.12a)$$

$$\frac{\partial f(W, \boldsymbol{\theta})}{\theta_l} = 2\left(\sum_{j=1}^{k} w_{ji}f_l(\mathbf{z}_j) - f_l(\mathbf{x}_i)\right) = 0 \qquad (8.12b)$$

for all $j = 1, 2, \ldots, k$ and $l = 0, 1, \ldots, J$.

Equation (8.12) can be written in the following matrix form:

$$
\begin{pmatrix}
C_{11} & \cdots & C_{1k} & f_0(\mathbf{z}_1) & \cdots & f_J(\mathbf{z}_1) \\
C_{21} & \cdots & C_{2k} & f_0(\mathbf{z}_2) & \cdots & f_J(\mathbf{z}_2) \\
\vdots & \vdots & \vdots & \vdots & \ddots & \vdots \\
C_{k1} & \cdots & C_{kk} & f_0(\mathbf{z}_k) & \cdots & f_J(\mathbf{z}_k) \\
f_0(\mathbf{z}_1) & \cdots & f_0(\mathbf{z}_k) & 0 & \cdots & 0 \\
\vdots & \vdots & \vdots & \vdots & \ddots & \vdots \\
f_J(\mathbf{z}_1) & \cdots & f_J(\mathbf{z}_k) & 0 & \cdots & 0
\end{pmatrix}
\begin{pmatrix}
w_{1i} \\
w_{2i} \\
\vdots \\
w_{ki} \\
\theta_0 \\
\vdots \\
\theta_J
\end{pmatrix}
=
\begin{pmatrix}
D_{1i} \\
D_{2i} \\
\vdots \\
D_{ki} \\
f_0(\mathbf{x}_i) \\
\vdots \\
f_J(\mathbf{x}_i)
\end{pmatrix}
\quad (8.13)
$$

for $i = 1, 2, \ldots, n$. The weights can be solved numerically from Equation (8.13). Comparing Equation (7.11) and Equation (8.13), we see that ordinary kriging is a special case of universal kriging.

Just as in ordinary kriging, the covariances C_{jl} and D_{ji} are modeled by theoretical variograms. Chapter 7 introduced three variogram models: exponential, spherical, and Gaussian. See Chapter 7 for details.

8.2   Implementation

The universal kriging method shares several functions with the ordinary kriging method. In this section, we implement the main function for the universal kriging method. Other functions are borrowed from Chapter 7.

The main function for the universal kriging method is implemented as follows:

```
1   ukrig <- function(Z, FZ, y, X, FX, varmodel) {
2       # Perform ordinary kriging prediction
3       #
4       # Args:
5       #   Z: a kxd matrix
6       #   FZ: a kxJ matrix to capture trend for Z
7       #   y: a vector of length k
8       #   X: a nxd matrix
9       #   FX: a nxJ matrix to capture trend for X
10      #   varmodel: a variogram model
11      #
12      # Returns:
13      #   a vector of predicted values for X
14
15      k <- nrow(Z)
16      n <- nrow(X)
17      d <- ncol(Z)
18      J <- ncol(FZ)
```

```
19
20    # calculate distance matrix for Z
21    hZ <- matrix(0, nrow=k, ncol=k)
22    for(i in 1:k) {
23      hZ[i,] <- (apply((Z - matrix(Z[i,], nrow=k, ncol=d,
          byrow=T))^2, 1, sum))^0.5
24    }
25
26    # calculate distance matrix between Z and X
27    hD <- matrix(0, nrow=k, ncol=n)
28    for(i in 1:k) {
29      hD[i,] <- (apply((X - matrix(Z[i,], nrow=n, ncol=d,
          byrow=T))^2, 1, sum))^0.5
30    }
31
32    # construct kriging equation system
33    V <- matrix(0, nrow=k+J, ncol=k+J)
34    V[1:k, 1:k] <- varmodel(hZ)
35    V[1:k, (k+1):(k+J)] <- FZ
36    V[(k+1):(k+J), 1:k] <- t(FZ)
37
38    D <- matrix(1, nrow=k+J, ncol=n)
39    D[1:k,] <- varmodel(hD)
40    D[(k+1):(k+J),] <- t(FX)
41
42    # solve equation
43    mW <- solve(V, D)
44
45    # perform prediction
46    mY <- matrix(0, nrow=k+J, ncol=1)
47    mY[1:k,1] <- y
48    yhat <- t(mW) %*% mY
49
50    return(yhat)
51  }
```

The function is named ukrig. It is similar to the okrig function implemented in Chapter 7. It has two more arguments than the okrig function: FZ and FX, which represent the mean function components.

The functions for variogram models and the function for fitting variogram models are implemented in Chapter 7. See that chapter for the code of these functions.

8.3 Applications

In this section, we apply the universal kriging method with various combinations of variogram models and experimental design methods to predict the fair market values of the synthetic dataset described in Appendix A.

We load and prepare the data as follows:

```
 1 > inforce <- read.csv("inforce.csv")
 2 >
 3 > vNames <- c("gbAmt", "gmwbBalance", "withdrawal",
     paste("FundValue", 1:10, sep=""))
 4 >
 5 > age <- with(inforce, (currentDate-birthDate)/365)
 6 > ttm <- with(inforce, (matDate - currentDate)/365)
 7 >
 8 > datN <- cbind(inforce[,vNames], data.frame(age=age,
     ttm=ttm))
 9 > datC <- inforce[,c("gender", "productType")]
10 > dat <- cbind(datN, datC)
11 >
12 > greek <- read.csv("Greek.csv")
13 > greek <- greek[order(greek$recordID),]
14 >
15 > # prepare data
16 > X <- model.matrix( ~ ., data=dat)[,-1]
17 > colnames(X)
18  [1] "gbAmt"            "gmwbBalance"      "withdrawal"
19  [4] "FundValue1"       "FundValue2"       "FundValue3"
20  [7] "FundValue4"       "FundValue5"       "FundValue6"
21 [10] "FundValue7"       "FundValue8"       "FundValue9"
22 [13] "FundValue10"      "age"              "ttm"
23 [16] "genderM"          "productTypeABRU" "
     productTypeABSU"
24 [19] "productTypeDBAB" "productTypeDBIB" "
     productTypeDBMB"
25 [22] "productTypeDBRP" "productTypeDBRU" "
     productTypeDBSU"
26 [25] "productTypeDBWB" "productTypeIBRP" "
     productTypeIBRU"
27 [28] "productTypeIBSU" "productTypeMBRP" "
     productTypeMBRU"
28 [31] "productTypeMBSU" "productTypeWBRP" "
     productTypeWBRU"
29 [34] "productTypeWBSU"
30 >
31 > vMin <- apply(X, 2, min)
32 > vMax <- apply(X, 2, max)
```

```
33 > X <- (X - matrix(vMin, nrow=nrow(X), ncol= ncol(X),
        byrow=TRUE)) / matrix(vMax-vMin, nrow=nrow(X), ncol=
        ncol(X), byrow=TRUE)
34 >
35 > FX <- cbind(1, X)
36 > head(FX, n=2)
37       gbAmt gmwbBalance withdrawal FundValue1
38 1 1 0.0400932           0          0 0.00000000
39 2 1 0.1187522           0          0 0.02786741
40   FundValue2 FundValue3 FundValue4 FundValue5 FundValue6
41 1          0 0.00000000 0.09300564 0.00000000 0.00000000
42 2          0 0.03977072 0.03454873 0.03903278 0.02955328
43   FundValue7 FundValue8 FundValue9 FundValue10        age
44 1 0.00000000          0 0.00000000  0.00000000 0.4288852
45 2 0.03841501          0 0.03611252  0.03653121 0.7771371
46        ttm genderM productTypeABRU productTypeABSU
47 1 0.6686936       0               0               0
48 2 0.6298548       1               0               0
49   productTypeDBAB productTypeDBIB productTypeDBMB
50 1               0               0               0
51 2               0               0               0
52   productTypeDBRP productTypeDBRU productTypeDBSU
53 1               0               0               0
54 2               0               0               0
55   productTypeDBWB productTypeIBRP productTypeIBRU
56 1               0               0               0
57 2               0               0               0
58   productTypeIBSU productTypeMBRP productTypeMBRU
59 1               0               0               0
60 2               0               0               0
61   productTypeMBSU productTypeWBRP productTypeWBRU
62 1               0               0               0
63 2               0               0               0
64   productTypeWBSU
65 1               0
66 2               0
```

In the above code, **FX** contains the values of the basis functions for modeling the mean function. The basis functions we used as

$$f_0(\mathbf{x}_i) = 1, \quad f_l(\mathbf{x}_i) = x_{il}, \quad l = 1, 2, \ldots, J$$

where x_{il} is the lth component of $\mathbf{x}_i$ and J is the number of dimensions of $\mathbf{x}_i$.

8.3.1 Universal Kriging with Latin Hypercube Sampling

In this subsection, we test the universal kriging method when Latin hypercube sampling is used to select representative policies.

The following output shows the performance of universal kriging with the exponential variogram model:

```
 1 > S <- read.table("lhs.csv", sep=",")
 2 > S <- S[,2]
 3 > Z <- X[S,]
 4 > FZ <- FX[S,]
 5 > y <- greek$fmv[S]/1000
 6 >
 7 > # fit variogram
 8 > {t1 <- proc.time()
 9 + res <- fitVarModel(Z, y, expVM, 100)
10 + proc.time()-t1}
11    user   system elapsed
12    0.41     0.03    0.43
13 > res
14              a              b              c
15       3.974054  -15424.628992   75994.250100
16 >
17 > {t1 <- proc.time()
18 + yhat <- ukrig(Z, FZ, y, X, FX, function(h) {(expVM(h,
       res[1], res[2], res[3]))})
19 + proc.time()-t1}
20    user   system elapsed
21  250.71    14.29   269.81
22 > calMeasure(greek$fmv/1000, yhat)
23 $pe
24 [1] -0.08325183
25
26 $r2
27 [1] 0.3448124
28
29 >
30 > png("explhs.png", width=8, height=4, units="in", res
       =300)
31 > par(mfrow=c(1,2), mar=c(4,4,1,1))
32 > plot(greek$fmv/1000, yhat, xlab="FMV(MC)", ylab="FMV(
       UK)")
33 > abline(0,1)
34 > qqplot(greek$fmv/1000, yhat, xlab="FMV(MC)", ylab="FMV
       (UK)")
35 > abline(0,1)
36 > dev.off()
37 null device
38              1
```

The percentage error is about -8.3% and the R^2 is about 0.34. These validation measures indicate that the results are not that good because the percentage error is relatively large and the R^2 is relatively low. The scatter and

QQ plots are shown in Figure 8.1. From the plots, we see that the model does not fit the left tail well. The universal kriging method underestimates the fair market values.

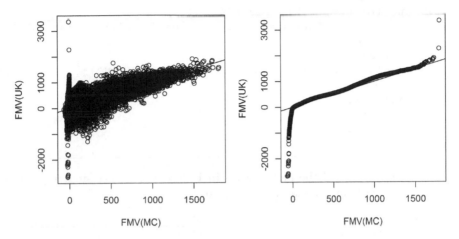

FIGURE 8.1: A scatter plot and a QQ plot of the fair market values calculated by Monte Carlo simulation and those predicted by universal kriging with an exponential variogram. The representative policies are selected by Latin hypercube sampling.

The following output shows the performance of universal kriging with a spherical variogram:

```
 1 > {t1 <- proc.time()
 2 + res <- fitVarModel(Z, y, sphVM, 100)
 3 + proc.time()-t1}
 4    user   system elapsed
 5    0.42    0.36    0.88
 6 > res
 7            a             b             c
 8     2.238043 -12424.305722  58737.822929
 9 >
10 > {t1 <- proc.time()
11 + yhat <- ukrig(Z, FZ, y, X, FX, function(h) {(sphVM(h,
       res[1], res[2], res[3]))})
12 + proc.time()-t1}
13    user   system elapsed
14  258.20   14.81   276.76
15 > calMeasure(greek$fmv/1000, yhat)
16 $pe
17 [1] 0.1184298
18
19 $r2
20 [1] 0.6199207
```

```
21
22 >
23 > png("sphlhs.png", width=8, height=4, units="in", res
     =300)
24 > par(mfrow=c(1,2), mar=c(4,4,1,1))
25 > plot(greek$fmv/1000, yhat, xlab="FMV(MC)", ylab="FMV(
     UK)")
26 > abline(0,1)
27 > qqplot(greek$fmv/1000, yhat, xlab="FMV(MC)", ylab="FMV
     (UK)")
28 > abline(0,1)
29 > dev.off()
30 null device
31                1
```

From the output and the plots shown in Figure 8.2, we see similar patterns
as the case with an exponential variogram.

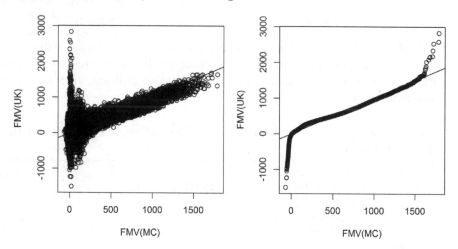

FIGURE 8.2: A scatter plot and a QQ plot of the fair market values calculated by Monte Carlo simulation and those predicted by universal kriging
with a spherical variogram. The representative policies are selected by Latin
hypercube sampling.

The following output shows the performance of universal kriging with a
Gaussian variogram:

```
1 > {t1 <- proc.time()
2 + res <- fitVarModel(Z, y, gauVM, 100)
3 + proc.time()-t1}
4    user   system elapsed
5    0.37     0.06    0.45
6 > res
7                a            b            c
```

```
 8        2.039751  -2563.415448 50597.212708
 9 >
10 > {t1 <- proc.time()
11 + yhat <- ukrig(Z, FZ, y, X, FX, function(h) {(gauVM(h,
       res[1], res[2], res[3]))})
12 + proc.time()-t1}
13     user   system elapsed
14   254.93    14.48   272.17
15 > calMeasure(greek$fmv/1000, yhat)
16 $pe
17 [1]  7.516662
18
19 $r2
20 [1]  -212.9296
21
22 >
23 > png("gaulhs.png", width=8, height=4, units="in", res
       =300)
24 > par(mfrow=c(1,2), mar=c(4,4,1,1))
25 > plot(greek$fmv/1000, yhat, xlab="FMV(MC)", ylab="FMV(
       UK)")
26 > abline(0,1)
27 > qqplot(greek$fmv/1000, yhat, xlab="FMV(MC)", ylab="FMV
       (UK)")
28 > abline(0,1)
29 > dev.off()
30 null device
31           1
```

The validation measures show that the model does not fit the fair market values well. The scatter plot and the QQ plot are shown in Figure 8.3. The plots show that the Gaussian variogram model caused the estimated value to be volatile. The predicted values have a wide range from −15,000 to 15,000.

8.3.2 Universal Kriging with Conditional Latin Hypercube Sampling

In this subsection, we illustrate the universal kriging method when conditional Latin hypercube sampling is used to select representative policies.

First, let us test universal kriging with an exponential variogram. The following output shows the test results:

```
1 > S <- read.table("clhs.csv", sep=",")
2 > S <- S[,2]
3 > Z <- X[S,]
4 > FZ <- FX[S,]
5 > y <- greek$fmv[S]/1000
6 >
```

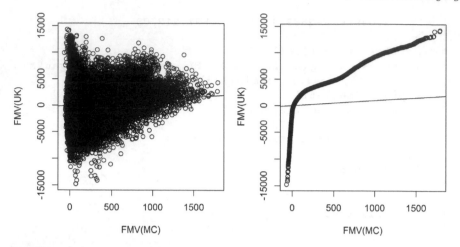

FIGURE 8.3: A scatter plot and a QQ plot of the fair market values calculated by Monte Carlo simulation and those predicted by universal kriging with a Gaussian variogram. The representative policies are selected by Latin hypercube sampling.

```
 7 > # fit variogram
 8 > {t1 <- proc.time()
 9 + res <- fitVarModel(Z, y, expVM, 100)
10 + proc.time()-t1}
11     user   system elapsed
12     0.41     0.06    0.47
13 > res
14          a            b            c
15     1.9733 -5061.9356 25822.8041
16 >
17 > {t1 <- proc.time()
18 + yhat <- ukrig(Z, FZ, y, X, FX, function(h) {(expVM(h,
      res[1], res[2], res[3]))})
19 + proc.time()-t1}
20     user   system elapsed
21   251.60    13.64   267.78
22 > calMeasure(greek$fmv/1000, yhat)
23 $pe
24 [1] 0.005445226
25
26 $r2
27 [1] 0.8329172
28
29 >
30 > png("expclhs.png", width=8, height=4, units="in", res
      =300)
31 > par(mfrow=c(1,2), mar=c(4,4,1,1))
```

```
32 > plot(greek$fmv/1000, yhat, xlab="FMV(MC)", ylab="FMV(
      UK)")
33 > abline(0,1)
34 > qqplot(greek$fmv/1000, yhat, xlab="FMV(MC)", ylab="FMV
      (UK)")
35 > abline(0,1)
36 > dev.off()
37 windows
38         2
```

The validation measures show that universal kriging with an exponential variogram and representative policies selected by conditional Latin hypercube sampling produced good results. The results are better than those produced by ordinary kriging with the same setting (see Section 7.3.2). The scatter and the QQ plots are shown in Figure 8.4. The plots show that the model does not fit the tails well.

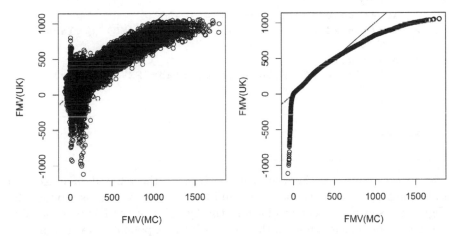

FIGURE 8.4: A scatter plot and a QQ plot of the fair market values calculated by Monte Carlo simulation and those predicted by universal kriging with an exponential variogram. The representative policies are selected by conditional Latin hypercube sampling.

The following output shows the performance of universal kriging with a spherical variogram:

```
1 > {t1 <- proc.time()
2 + res <- fitVarModel(Z, y, sphVM, 100)
3 + proc.time()-t1}
4     user   system elapsed
5     0.41    0.00    0.40
6 > res
7              a              b              c
```

```
 8        1.828089   -739.281594  20622.377412
 9 >
10 > {t1 <- proc.time()
11 + yhat <- ukrig(Z, FZ, y, X, FX, function(h) {(sphVM(h,
      res[1], res[2], res[3]))})
12 + proc.time()-t1}
13    user   system elapsed
14   255.70    15.15   273.75
15 > calMeasure(greek$fmv/1000, yhat)
16 $pe
17 [1] -0.01946165
18
19 $r2
20 [1] 0.9410583
21
22 >
23 > png("sphclhs.png", width=8, height=4, units="in", res
      =300)
24 > par(mfrow=c(1,2), mar=c(4,4,1,1))
25 > plot(greek$fmv/1000, yhat, xlab="FMV(MC)", ylab="FMV(
      UK)")
26 > abline(0,1)
27 > qqplot(greek$fmv/1000, yhat, xlab="FMV(MC)", ylab="FMV
      (UK)")
28 > abline(0,1)
29 > dev.off()
30 windows
31        2
```

The output shows that the results are better than those based on an exponential variogram. The R^2 is above 0.94 and the percentage error is around -2%. Figure 8.5 shows the scatter plot and the QQ plot of the fair market values. We see a similar pattern as in the case with the exponential variogram.

The results based on a Gaussian variogram are shown in the following output:

```
 1 > {t1 <- proc.time()
 2 + res <- fitVarModel(Z, y, gauVM, 100)
 3 + proc.time()-t1}
 4    user   system elapsed
 5    0.42     0.07    0.49
 6 > res
 7              a              b              c
 8      1.539091   2304.095250  17525.789260
 9 >
10 > {t1 <- proc.time()
11 + yhat <- ukrig(Z, FZ, y, X, FX, function(h) {(gauVM(h,
      res[1], res[2], res[3]))})
12 + proc.time()-t1}
```

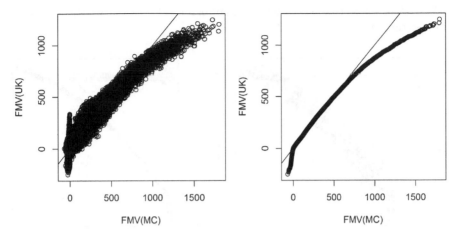

FIGURE 8.5: A scatter plot and a QQ plot of the fair market values calculated by Monte Carlo simulation and those predicted by universal kriging with a spherical variogram. The representative policies are selected by conditional Latin hypercube sampling.

```
13    user   system  elapsed
14   250.71   14.24   268.49
15 > calMeasure(greek$fmv/1000, yhat)
16 $pe
17 [1] -0.0177252
18
19 $r2
20 [1] 0.9432273
21
22 >
23 > png("gauclhs.png", width=8, height=4, units="in", res
       =300)
24 > par(mfrow=c(1,2), mar=c(4,4,1,1))
25 > plot(greek$fmv/1000, yhat, xlab="FMV(MC)", ylab="FMV(
       UK)")
26 > abline(0,1)
27 > qqplot(greek$fmv/1000, yhat, xlab="FMV(MC)", ylab="FMV
       (UK)")
28 > abline(0,1)
29 > dev.off()
30 windows
31         2
```

The validation measures show that using a Gaussian variogram produced accurate results in this case. The scatter plot and the QQ plot are shown in

Figure 8.6. From the figure, we see that using a Gaussian variogram did not fit the tails well.

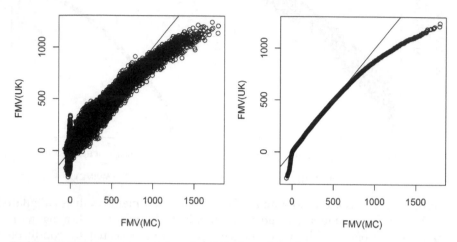

FIGURE 8.6: A scatter plot and a QQ plot of the fair market values calculated by Monte Carlo simulation and those predicted by universal kriging with a Gaussian variogram. The representative policies are selected by conditional Latin hypercube sampling.

8.3.3   Universal Kriging with Hierarchical k-Means

In this section, we present test results of universal kriging when hierarchical k-means is used to select representative policies.

The following output shows the results when the exponential variogram was used:

```
 1 > S <- read.table("hkmeans.csv", sep=",")
 2 > S <- S[,2]
 3 > Z <- X[S,]
 4 > FZ <- FX[S,]
 5 > y <- greek$fmv[S]/1000
 6 >
 7 > # fit variogram
 8 > {t1 <- proc.time()
 9 + res <- fitVarModel(Z, y, expVM, 100)
10 + proc.time()-t1}
11     user   system  elapsed
12     0.47     0.38     1.09
13 > res
14              a              b              c
15     3.394879  -3908.378835  34032.994913
16 >
```

```
17 > {t1 <- proc.time()
18 + yhat <- ukrig(Z, FZ, y, X, FX, function(h) {(expVM(h,
       res[1], res[2], res[3]))})
19 + proc.time()-t1}
20    user  system elapsed
21  250.13    7.12  260.60
22 > calMeasure(greek$fmv/1000, yhat)
23 $pe
24 [1] 0.000567231
25
26 $r2
27 [1] 0.9115645
28
29 >
30 > png("exphkmeans.png", width=8, height=4, units="in",
       res=300)
31 > par(mfrow=c(1,2), mar=c(4,4,1,1))
32 > plot(greek$fmv/1000, yhat, xlab="FMV(MC)", ylab="FMV(
       UK)")
33 > abline(0,1)
34 > qqplot(greek$fmv/1000, yhat, xlab="FMV(MC)", ylab="FMV
       (UK)")
35 > abline(0,1)
36 > dev.off()
37 windows
38       2
```

The output shows that the percentage error is around 0.06%, which means
that the total fair market value predicted by the model is very close to that
calculated by Monte Carlo simulation. The R^2 is also above 0.9. Figure 8.7
shows the scatter plot and the QQ plot of the fair market values. The plots
show that the model does not fit the tails well.

To see the performance of universal kriging with a spherical variogram, we
proceed as follows:

```
 1 > {t1 <- proc.time()
 2 + res <- fitVarModel(Z, y, sphVM, 100)
 3 + proc.time()-t1}
 4    user  system elapsed
 5    0.39    0.02    0.40
 6 > res
 7           a             b              c
 8    1.968632 -1710.723617 25409.565446
 9 >
10 > {t1 <- proc.time()
11 + yhat <- ukrig(Z, FZ, y, X, FX, function(h) {(sphVM(h,
       res[1], res[2], res[3]))})
12 + proc.time()-t1}
13    user  system elapsed
```

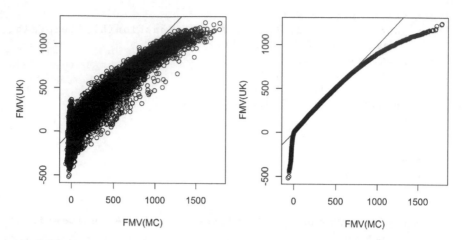

FIGURE 8.7: A scatter plot and a QQ plot of the fair market values calculated by Monte Carlo simulation and those predicted by universal kriging with an exponential variogram. The representative policies are selected by hierarchical k-means.

```
14  247.46    13.62    262.69
15 > calMeasure(greek$fmv/1000, yhat)
16 $pe
17 [1] -0.001190777
18
19 $r2
20 [1] 0.9091358
21
22 >
23 > png("sphhkmeans.png", width=8, height=4, units="in",
       res=300)
24 > par(mfrow=c(1,2), mar=c(4,4,1,1))
25 > plot(greek$fmv/1000, yhat, xlab="FMV(MC)", ylab="FMV(
       UK)")
26 > abline(0,1)
27 > qqplot(greek$fmv/1000, yhat, xlab="FMV(MC)", ylab="FMV
       (UK)")
28 > abline(0,1)
29 > dev.off()
30 windows
31       2
```

The validation measures show that the performance with a spherical variogram is similar to that with an exponential variogram. The scatter plot and the QQ plot are shown in Figure 8.8. We see similar patterns as in Figure 8.7.

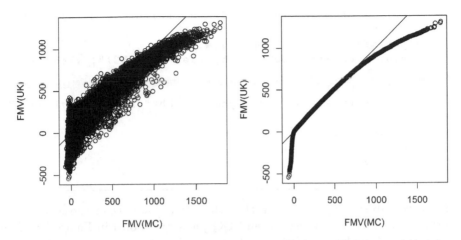

FIGURE 8.8: A scatter plot and a QQ plot of the fair market values calculated by Monte Carlo simulation and those predicted by universal kriging with a spherical variogram. The representative policies are selected by hierarchical *k*-means.

The following output shows the results when the Gaussian variogram model was used:

```
 1 > {t1 <- proc.time()
 2 + res <- fitVarModel(Z, y, gauVM, 100)
 3 + proc.time()-t1}
 4    user   system elapsed
 5    0.40    0.11    0.52
 6 > res
 7             a             b            c
 8      1.927708    3598.391784 21513.017652
 9 >
10 > {t1 <- proc.time()
11 + yhat <- ukrig(Z, FZ, y, X, FX, function(h) {(gauVM(h,
      res[1], res[2], res[3]))})
12 + proc.time()-t1}
13    user   system elapsed
14  257.68    14.10   278.06
15 > calMeasure(greek$fmv/1000, yhat)
16 $pe
17 [1] -0.005930386
18
19 $r2
20 [1] 0.8668875
21
22 >
```

```
23 > png("gauhkmeans.png", width=8, height=4, units="in",
      res=300)
24 > par(mfrow=c(1,2), mar=c(4,4,1,1))
25 > plot(greek$fmv/1000, yhat, xlab="FMV(MC)", ylab="FMV(
      UK)")
26 > abline(0,1)
27 > qqplot(greek$fmv/1000, yhat, xlab="FMV(MC)", ylab="FMV
      (UK)")
28 > abline(0,1)
29 > dev.off()
30 null device
31            1
```

From the above output, we see that the R^2 is lower than those from the previous two cases. The scatter and QQ plots are shown in Figure 8.9. The patterns are similar as before.

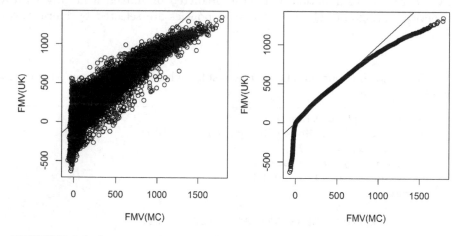

FIGURE 8.9: A scatter plot and a QQ plot of the fair market values calculated by Monte Carlo simulation and those predicted by universal kriging with a Gaussian variogram. The representative policies are selected by hierarchical k-means.

8.4   Summary

In this chapter, we introduced and implemented the universal kriging method, which allows for a smoothly varying and nonstationary trend. We also tested the universal kriging method with different combinations of variogram models

TABLE 8.1: Validation measures of ordinary kriging with different combinations of variogram models and experimental design methods.

	PE			R^2		
	LHS	cLHS	Hkmeans	LHS	cLHS	Hkmeans
Exponential	−8.33%	0.54%	0.06%	0.3448	0.8329	0.9116
Spherical	11.84%	−1.95%	−0.12%	0.6199	0.9411	0.9091
Gaussian	751.67%	−1.77%	−0.59%	−212.9296	0.9432	0.8669

and experimental design methods. The results are summarized in Table 8.1. The validation measures in the table show that hierarchical k-means and conditional Latin hypercube sampling are superior. When hierarchical k-means and conditional Latin hypercube sampling were used to select representative policies, we obtained results with low percentage errors and high R^2 no matter what variogram models were used.

For more information about universal kriging, readers are referred to Cressie (1993) and Wackernagel (2003).

9

GB2 Regression Model

As seen in Chapters 7 and 8, kriging methods do not fit the tails of the fair market values well, because the distribution of fair market values is highly skewed. In this chapter, we introduce a shifted GB2 (Generalized Beta of the second kind) regression model, which is able to capture the skewness of the distribution of the fair market values. In comparison to kriging methods, the GB2 regression model does not require calculating the distances between the variable annuity policies.

9.1 Description of the Model

The probability density function of a GB2 random variable, Z, is given by:

$$f(z) = \frac{|a|}{bB(p,q)} \left(\frac{z}{b}\right)^{ap-1} \left[1 + \left(\frac{z}{b}\right)^a\right]^{-p-q}, \quad z > 0, \tag{9.1}$$

where $a \neq 0$, $p > 0$, $q > 0$ are shape parameters, $b > 0$ is the scale parameter, and $B(p,q)$ is the Beta function. The expectation of Z exists when $-p < \frac{1}{a} < q$ and is given by

$$E[Z] = \frac{bB\left(p + \frac{1}{a}, q - \frac{1}{a}\right)}{B(p,q)}. \tag{9.2}$$

In the GB2 regression model, the fair market value is modeled as a shifted GB2 distribution. That is, the fair market value Y is modeled as:

$$Y = Z - c, \tag{9.3}$$

where Z follows a GB2 distribution with four parameters (a, b, p, q) and c is a location parameter, which provides the flexibility to shift the distribution to allow for negative fair market values.

Let $\mathbf{z}_i = (1, z_{i1}, z_{i2}, \ldots, z_{ik})'$ be a vector containing the covariate values of the ith representative variable annuity policy and let $\boldsymbol{\beta} = (\beta_0, \beta_1, \ldots, \beta_k)'$ be a vector of regression coefficients. Note that in kriging methods, covariates are referred to as locations. The covariates are incorporated into the regression model through the scale parameter, i.e.,

$$b(\mathbf{z}_i) = \exp(\mathbf{z}_i'\boldsymbol{\beta}).$$

117

It is possible to incorporate covariates through the shape parameters, but this poses additional challenges. See Gan and Valdez (2018c) for details.

The method of maximum likelihood is used to estimate the parameters. Since the covariates are incorporated through the scale parameter $b(\mathbf{z}_i) = \exp(\mathbf{z}_i'\boldsymbol{\beta})$, the log-likelihood function of the model is given below (Gan and Valdez, 2018c):

$$L(a,p,q,c,\boldsymbol{\beta}) = k \ln \frac{|a|}{B(p,q)} - ap \sum_{i=1}^{k} \mathbf{z}_i'\boldsymbol{\beta} + (ap-1) \sum_{i=1}^{k} \ln(v_i + c) -$$

$$(p+q) \sum_{i=1}^{k} \ln \left[1 + \left(\frac{v_i + c}{\exp(\mathbf{z}_i'\boldsymbol{\beta})} \right)^a \right], \tag{9.4}$$

where k is the number of representative policies and v_i denotes the fair market value of the ith representative policy. As with any metamodels, the representative policies are used to calibrate the model.

Once the parameters for the GB2 model are estimated from the data, the expectation given in Equation (9.2) is used to predict the fair market value of an arbitrary policy in the portfolio. Mathematically, the prediction is calculated as follows:

$$\widehat{y}_i = \frac{\exp(\mathbf{x}_i'\boldsymbol{\beta}) B \left(p + \frac{1}{a}, q - \frac{1}{a} \right)}{B(p,q)} - c, \quad i = 1,2,\ldots,n, \tag{9.5}$$

where $a,p,q,c,\boldsymbol{\beta}$ are parameters estimated from the data.

Estimating the parameters is quite challenging because the log-likelihood function can have multiple local maxima and the optimization algorithm can return a suboptimal set of parameters depending on the initial set of parameters. Multi-stage optimization is used to estimate the parameters of the GB2 regression model. The purpose of multi-stage optimization is to find a good initial set of parameters.

In the first stage, the shift parameter and the regression coefficients are fixed and the optimum shape parameters are fitted to the data. In this stage, the following profile log-likelihood function is maximized:

$$L_1(a,p,q) = L(a,p,q,c_0,\boldsymbol{\beta}_0), \tag{9.6}$$

where $c_0 = -\min\{v_1,\ldots,v_s\} + 10^{-6}$ and

$$\boldsymbol{\beta}_0 = \left(\ln \left(\frac{1}{s} \sum_{i=1}^{s} v_i + c_0 \right), 0, 0, \ldots, 0 \right)'. \tag{9.7}$$

In the second stage, the optimum shift parameter is fitted by fixing the shape parameters and the regression coefficients. In this stage, the following profile log-likelihood function is optimized:

$$L_2(c) = L(\widehat{a},\widehat{p},\widehat{q},c,\boldsymbol{\beta}_0), \tag{9.8}$$

where $\widehat{a}, \widehat{p}$, and $\widehat{q}$ are obtained from the first stage and $\boldsymbol{\beta}_0$ is defined in Equation (9.7).

In the third stage, the optimum regression coefficients are estimated by fixing the shape parameters and the shift parameter. In this stage, the following profile log-likelihood function is maximized:

$$L_3(\boldsymbol{\beta}) = L(\widehat{a}, \widehat{p}, \widehat{q}, \widehat{c}, \boldsymbol{\beta}), \tag{9.9}$$

where $\widehat{a}, \widehat{p}$, and $\widehat{q}$ are obtained from the first stage and $\widehat{c}$ is obtained from the second stage. The following initial set of parameters is used:

$$\boldsymbol{\beta}_1 = \left(\ln \left(\frac{1}{s} \sum_{i=1}^{s} v_i + \widehat{c} \right), 0, 0, \ldots, 0 \right)', \tag{9.10}$$

where $\widehat{c}$ is obtained from the second stage. Note that $\boldsymbol{\beta}_1$ is slightly different from $\boldsymbol{\beta}_0$ defined in Equation (9.7) in that c_0 is changed to $\widehat{c}$.

In the last stage, a full maximization of the log-likelihood function is performed with the following set of initial values:

$$\boldsymbol{\theta}_0 = (\widehat{a}, \widehat{p}, \widehat{q}, \widehat{c}, \widehat{\boldsymbol{\beta}}), \tag{9.11}$$

where $\widehat{a}, \widehat{p}$, and $\widehat{q}$ are obtained from the first stage, $\widehat{c}$ is obtained from the second stage, and $\widehat{\boldsymbol{\beta}}$ is obtained from the third stage.

9.2 Implementation

In this section, we implement the GB2 regression model for predicting fair market values of variable annuity policies. In particular, we implement the maximum likelihood method for parameter estimation with the multi-stage optimization method.

Before we implement the log-likelihood function, let us create functions of the GB2 distribution. The following R code implements the probability density function of the GB2 distribution:

```
1  dGB2 <- function(x,a,b,gamma1,gamma2) {
2    num <- abs(a)*(x^(a*gamma1-1))*(b^(a*gamma2))
3    temp <- (b^a+x^a)^(gamma1+gamma2)
4    den <- beta(gamma1,gamma2)*temp
5    result <- num/den
6    return(result)
7  }
```

The function is named dGB2 and will be used to calculate the log-likelihood function. To make predictions, we also need a function to calculate the expectation defined in Equation (9.2). The expectation is calculated by the following R function:

```
1 eGB2 <- function(a, b, p, q) {
2   return( b * beta(p + 1/a, q - 1/a) / beta(p, q) )
3 }
```

In the two functions given above, we used the R function `beta` to calculate the Beta function.

The following three R functions implement the profile log-likelihood functions L_1, L_2, and L_3 defined in Equations (9.6), (9.8), and (9.9), respectively:

```
1 negllS3a <- function(param, y) {
2   c <- -min(y) + 1e-6
3   beta0 <- log(mean(y + c))
4   temp <- log(dGB2( y+ c, a= param[1], b=exp(beta0),
        gamma1= param[2], gamma2= param[3]))
5   result <- -sum(temp)
6   if(is.nan(result) || abs(result) > 1e+10) {
7     result = 1e+10
8   }
9   return(result)
10 }
11
12 negllS3b <- function(vp, y, param) {
13   c <- -min(y) + 1e-6
14   beta0 <- log(mean(y + c))
15   temp <- log(dGB2( y+ vp, a= param[1], b=exp(beta0),
        gamma1= param[2], gamma2= param[3]))
16   result <- -sum(temp)
17   if(is.nan(result) || abs(result) > 1e+10) {
18     result = 1e+10
19   }
20   return(result)
21 }
22
23 negllS3c <- function(vp, X, y, param1) {
24   param <- c(param1, vp)
25   xbeta <- X %*% as.matrix(param[-c(1:4)], ncol=1)
26
27   temp <- log(dGB2( y+ param[4], a= param[1], b=exp(
        xbeta), gamma1= param[2], gamma2= param[3]))
28   result <- -sum(temp)
29   if(is.nan(result) || abs(result) > 1e+10) {
30     result = 1e+10
31   }
32   return(result)
33 }
```

In the above functions, we actually calculate the negative log-likelihood functions. Minimizing the negative log-likelihood function is equivalent to maximizing the log-likelihood function. We also use a numerical strategy in these functions. That is, when the value is invalid (e.g., the parameters are out of the range) or the value has a large magnitude, we assign a large value (10^{10}) to the value. This strategy will help the optimization algorithm to look for parameter values in valid ranges.

The following R function implements the negative log-likelihood function used in the last stage of the optimization process:

```
1  negllS <- function(param, X, y) {
2    xbeta <- X %*% as.matrix(param[-c(1:4)], ncol=1)
3
4    temp <- log(dGB2( y+ param[4], a= param[1], b=exp(
         xbeta), gamma1= param[2], gamma2= param[3]))
5    result <- -sum(temp)
6    if(is.nan(result) || abs(result) > 1e+10 ) {
7      result = 1e+10
8    }
9    return(result)
10 }
```

Fitting a GB2 regression model to the data is implemented in the following R function:

```
1  gb2 <- function(X, y, S) {
2    # GB2 regression model
3    #
4    # args:
5    #   X: nxd design matrix of the whole inforce
6    #   y: a vector of k fair market values of
          representative policies
7    #   S: a vector of k indices
8    #
9    # returns:
10   #   a vector of predicted values
11
12   L <- c(0, 0, 0)
13   U <- c(10, 10,  10)
14   NS <- 100
15   SP <- matrix(0, nrow=NS, ncol=length(L))
16   for(k in 1:length(L)) {
17     SP[, k] <- runif(NS, min=L[k], max=U[k])
18   }
19
20   vLL <- matrix(0, nrow=NS, ncol=1)
21   for(i in 1:NS) {
22     vLL[i] <- negllS3a(SP[i,], y=y)
23   }
24   SP1 <- SP[order(vLL),]
25   SP <- SP1[1:10,]
```

```
26
27    # Stage 1
28    mRes <- matrix(0, nrow=nrow(SP), ncol=2+length(L))
29    for(i in 1:nrow(SP)) {
30      fit.GB2 <- optim(SP[i,], negllS3a, NULL, y=y,
                control = list(maxit =10000))
31      mRes[i, 1] <- fit.GB2$value
32      mRes[i, 2] <- fit.GB2$convergence
33      mRes[i, -c(1:2)] <- fit.GB2$par
34    }
35
36    # Stage 2
37    iMin <- which.min(mRes[,1])
38    ahat <- mRes[iMin, 3]
39    phat <- mRes[iMin, 4]
40    qhat <- mRes[iMin, 5]
41    fit2 <- optimize(negllS3b, interval=c(-min(y)+1e-6,
                -10*min(y)), y=y, param=mRes[iMin, 3:5])
42    chat <- fit2$minimum
43
44    # Stage 3
45    fit3 <- optim(c(log(mean(y) + chat), rep(0, ncol(X)-1)
                ), negllS3c, NULL, X=X[S,], y=y,
46      param1=c(mRes[iMin, 3:5], chat), control = list(
                maxit =50000))
47
48    # Stage 4
49    fit4 <- optim(c(mRes[iMin, 3:5], chat, fit3$par),
                negllS, NULL, X=X[S,], y=y,
50      control = list(maxit =50000))
51
52    param.hat <- fit4$par
53    a <- param.hat[1]
54    p <- param.hat[2]
55    q <- param.hat[3]
56    c <- param.hat[4]
57    b <- exp(X %*% as.matrix(param.hat[-c(1:4)], ncol=1))
58    print(round(fit4$par,4))
59    yhat <- eGB2(a, b, p, q) - c
60
61    return(yhat)
62 }
```

The function is called gb2 and implements the four-stage optimization process described in the previous section. It requires three arguments, which are explained in the comments of this function.

The first stage needs some further explanation. In this stage, we maximize $L_1(a, p, q)$ by using the R function optim to minimize $-L_1(a, p, q)$. The function `optim` requires an initial set of values for a, p, and q. We generate $N = 100$ sets of initial values (a, p, q) randomly from the parameter space $(0.1, 10)^3$ to see the sensitivity of the initial values. To reduce the computation, we first calculate the profile log-likelihood function L_1 at these sets of initial values and select 10 sets that produce the highest L_1 values. Then we apply `optim` with these 10 sets of initial values and look at the results from the 10 runs.

9.3 Applications

In this section, we apply the GB2 regression model to predict the fair market values of the synthetic dataset described in Appendix A.

Before we fit the models, we first load and prepare the data as follows:

```
 1 > inforce <- read.csv("inforce.csv")
 2 >
 3 > vNames <- c("gbAmt", "gmwbBalance", "withdrawal",
       paste("FundValue", 1:10, sep=""))
 4 >
 5 > age <- with(inforce, (currentDate-birthDate)/365)
 6 > ttm <- with(inforce, (matDate - currentDate)/365)
 7 >
 8 > datN <- cbind(inforce[,vNames], data.frame(age=age,
       ttm=ttm))
 9 > datC <- inforce[,c("gender", "productType")]
10 >
11 > dat <- cbind(datN, datC)
12 >
13 > greek <- read.csv("Greek.csv")
14 > greek <- greek[order(greek$recordID),]
15 >
16 > # prepare data
17 > X <- model.matrix( ~ ., data=dat)[,-1]
18 > colnames(X)
19  [1] "gbAmt"              "gmwbBalance"        "withdrawal"
20  [4] "FundValue1"         "FundValue2"         "FundValue3"
21  [7] "FundValue4"         "FundValue5"         "FundValue6"
22 [10] "FundValue7"         "FundValue8"         "FundValue9"
23 [13] "FundValue10"        "age"                "ttm"
24 [16] "genderM"            "productTypeABRU"    "
       productTypeABSU"
25 [19] "productTypeDBAB"    "productTypeDBIB"    "
       productTypeDBMB"
26 [22] "productTypeDBRP"    "productTypeDBRU"    "
       productTypeDBSU"
```

```
27 [25] "productTypeDBWB" "productTypeIBRP" "
      productTypeIBRU"
28 [28] "productTypeIBSU" "productTypeMBRP" "
      productTypeMBRU"
29 [31] "productTypeMBSU" "productTypeWBRP" "
      productTypeWBRU"
30 [34] "productTypeWBSU"
31 >
32 > vMin <- apply(X, 2, min)
33 > vMax <- apply(X, 2, max)
34 > X <- (X - matrix(vMin, nrow=nrow(X), ncol= ncol(X),
      byrow=TRUE)) / matrix(vMax-vMin, nrow=nrow(X), ncol=
      ncol(X), byrow=TRUE)
35 > X <- cbind(1, X) # add an intercept
```

In the above output, X is a design matrix containing the covariate information. In Line 35, we also add an intercept to the design matrix.

9.3.1 GB2 with Latin Hypercube Sampling

To fit a GB2 model based on representative policies selected by Latin hypercube sampling, we proceed as follows:

```
1 > S <- read.table("lhs.csv", sep=",")
2 > S <- S[,2]
3 >
4 > y <- greek$fmv[S]/1000
5 >
6 > {t1 <- proc.time()
7 + set.seed(1)
8 + yhat <- gb2(X, y, S)
9 + proc.time()-t1}
10  [1] -2.9189  2.2160  1.2223 43.1040  4.4773  1.2417
11  [7]  0.0023 -0.3145 -0.7349 -0.5408 -0.1436  0.1556
12 [13] -0.3755 -0.6718 -0.5975 -0.4115 -0.0899 -0.3966
13 [19] -0.1058 -1.1960 -0.0451  2.4732  0.7248  0.8819
14 [25]  1.0345  0.8356 -0.3671 -0.9124 -0.8548 -0.1260
15 [31] -0.0109  2.1084  1.4079 -0.2506  1.8454  0.6228
16 [37]  0.0669  0.2327  0.1502
17    user  system elapsed
18   13.76    0.03   13.82
19 Warning messages:
20 1: In beta(gamma1, gamma2) : NaNs produced
21 2: In beta(gamma1, gamma2) : NaNs produced
22 3: In beta(gamma1, gamma2) : NaNs produced
23 4: In beta(gamma1, gamma2) : NaNs produced
24 5: In beta(gamma1, gamma2) : NaNs produced
25 6: In beta(gamma1, gamma2) : NaNs produced
26 7: In beta(gamma1, gamma2) : NaNs produced
```

```
27  8: In beta(gamma1, gamma2) : NaNs produced
28  9: In beta(gamma1, gamma2) : NaNs produced
29  > calMeasure(greek$fmv/1000, yhat)
30  $pe
31  [1]  -0.03164181
32
33  $r2
34  [1]  0.798557
```

In the above output, we first read the indices of the representative policies that are produced in Chapter 4. It took the function about 14 seconds to fit the GB2 model. The fitted regression coefficients are shown in Lines 10–16. We also see some warnings from the `beta` function. These warnings were caused by the invalid arguments from the optimization algorithm.

To visualize the goodness-of-fit, we produce a scatter plot and a QQ plot as follows:

```
1  > png("gb2lhs.png", width=8, height=4, units="in", res
      =300)
2  > par(mfrow=c(1,2), mar=c(4,4,1,1))
3  > plot(greek$fmv/1000, yhat, xlab="FMV(MC)", ylab="FMV(
      GB2)")
4  > abline(0,1)
5  > qqplot(greek$fmv/1000, yhat, xlab="FMV(MC)", ylab="FMV
      (GB2)")
6  > abline(0,1)
7  > dev.off()
8  null device
9            1
```

The resulting plots are shown in Figure 9.1. The QQ plot shows that the GB2 regression model has a good overall fit to the data. The model fit the left tail well, while the fit at the right tail is a little bit off.

9.3.2 GB2 with Conditional Latin Hypercube Sampling

To see the performance of the GB2 regression model when conditional Latin hypercube sampling is used to select representative policies, we proceed as follows:

```
1  > S <- read.table("clhs.csv", sep=",")
2  > S <- S[,2]
3  > y <- greek$fmv[S]/1000
4  > {t1 <- proc.time()
5  + set.seed(1)
6  + yhat <- gb2(X, y, S)
7  + proc.time()-t1}
8    [1]   5.9325   0.8240   2.0630  43.8082   4.6425   2.9993
```

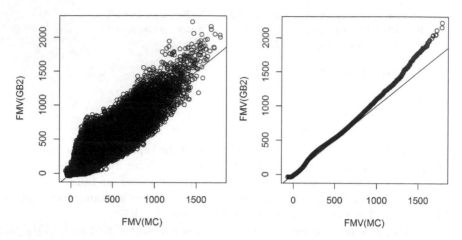

FIGURE 9.1: A scatter plot and a QQ plot of the fair market values calculated by Monte Carlo simulation and those predicted by GB2 regression. The representative policies are selected by Latin hypercube sampling.

```
 9  [7]  -0.0983  -0.8456  -1.5055  -1.8865  -1.9306  -1.6441
10 [13]  -2.3276  -1.0611  -0.8455  -0.8612  -0.6972  -1.6070
11 [19]  -0.1439  -0.7369  -0.0780   1.6582   0.3925   0.4478
12 [25]   0.4993   0.2096  -0.4642  -0.8045  -0.6429   0.1356
13 [31]   0.1499   1.2886   0.7925  -0.0517   1.3383   0.2342
14 [37]   0.2425   0.0922   0.2207
15    user  system elapsed
16   16.35    0.00   16.36
17 Warning messages:
18 1: In beta(gamma1, gamma2) : NaNs produced
19 2: In beta(gamma1, gamma2) : NaNs produced
20 3: In beta(gamma1, gamma2) : NaNs produced
21 4: In beta(gamma1, gamma2) : NaNs produced
22 5: In beta(gamma1, gamma2) : NaNs produced
23 6: In beta(gamma1, gamma2) : NaNs produced
24 7: In beta(gamma1, gamma2) : NaNs produced
25 8: In beta(gamma1, gamma2) : NaNs produced
26 > calMeasure(greek$fmv/1000, yhat)
27 $pe
28 [1] 0.1011085
29
30 $r2
31 [1] 0.6595339
32
33 >
34 > png("gb2clhs.png", width=8, height=4, units="in", res
      =300)
35 > par(mfrow=c(1,2), mar=c(4,4,1,1))
```

```
36 > plot(greek$fmv/1000, yhat, xlab="FMV(MC)", ylab="FMV(
     GB2)")
37 > abline(0,1)
38 > qqplot(greek$fmv/1000, yhat, xlab="FMV(MC)", ylab="FMV
     (GB2)")
39 > abline(0,1)
40 > dev.off()
41 null device
42             1
```

From the above output, we see that we changed the indices of the representative policies. The validation measures indicate that results produced by the GB2 regression model with conditional Latin hypercube sampling are worse than those produced by the GB2 regression model with Latin hypercube sampling. The R^2 is about 0.66, which is lower than the R^2 from the previous subsection.

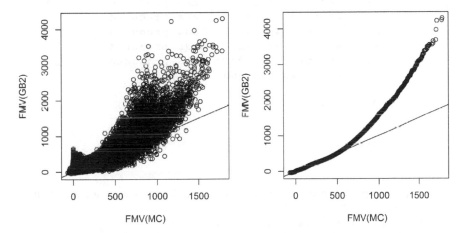

FIGURE 9.2: A scatter plot and a QQ plot of the fair market values calculated by Monte Carlo simulation and those predicted by GB2 regression. The representative policies are selected by conditional Latin hypercube sampling.

Figure 9.2 shows the scatter and the QQ plots. From the figure, we see that the GB2 regression model does not fit the right tail well. Comparing Figure 9.2 with Figure 9.1, we see that experimental designs have a big impact on the performance of GB2 regression models.

9.3.3 GB2 with Hierarchical k-Means

The following output shows the performance of the GB2 regression model when the hierarchical k-means is used to select representative policies:

```
1 > S <- read.table("hkmeans.csv", sep=",")
2 > S <- S[,2]
```

```
 3 > y <- greek$fmv[S]/1000
 4 > {t1 <- proc.time()
 5 + set.seed(1)
 6 + yhat <- gb2(X, y, S)
 7 + proc.time()-t1}
 8   [1]   6.2051   0.6868   1.4052 25.1646   4.2527   2.2432
 9   [7]  -0.6274  -0.1430  -0.9716 -2.0819  -4.2614  -2.9546
10  [13]   1.6524  -0.7847   0.4136  0.8807  -1.7987   0.3966
11  [19]  -0.0426  -0.9512  -0.0173  1.9439   0.4005   0.4674
12  [25]   0.5622   0.3564  -0.7637 -1.2881  -1.0647   0.5792
13  [31]   0.3824   1.7151   0.9621  0.0153   1.6102   0.2918
14  [37]   0.4697   0.3184   0.3917
15    user   system  elapsed
16   19.50     0.02    19.57
17 Warning messages:
18 1: In beta(gamma1, gamma2) : NaNs produced
19 2: In beta(gamma1, gamma2) : NaNs produced
20 3: In beta(gamma1, gamma2) : NaNs produced
21 4: In beta(gamma1, gamma2) : NaNs produced
22 5: In beta(gamma1, gamma2) : NaNs produced
23 6: In beta(gamma1, gamma2) : NaNs produced
24 7: In beta(gamma1, gamma2) : NaNs produced
25 > calMeasure(greek$fmv/1000, yhat)
26 $pe
27 [1] 0.09914421
28
29 $r2
30 [1] 0.4916871
31
32 >
33 > png("gb2hkmeans.png", width=8, height=4, units="in",
      res=300)
34 > par(mfrow=c(1,2), mar=c(4,4,1,1))
35 > plot(greek$fmv/1000, yhat, xlab="FMV(MC)", ylab="FMV(
      OK)")
36 > abline(0,1)
37 > qqplot(greek$fmv/1000, yhat, xlab="FMV(MC)", ylab="FMV
      (OK)")
38 > abline(0,1)
39 > dev.off()
40 null device
41          1
```

The validation measures in the output show that the GB2 regression model with representative policies selected by hierarchical k-means produced the worst results. The R^2 is around 0.49, which is lower than the previous two cases.

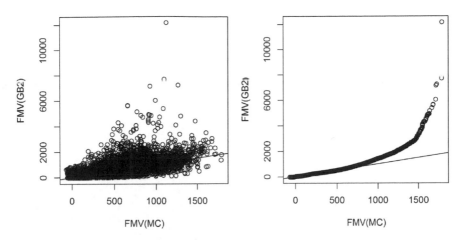

FIGURE 9.3: A scatter plot and a QQ plot of the fair market values calculated by Monte Carlo simulation and those predicted by GB2 regression. The representative policies are selected by hierarchical k-means.

Figure 9.3 shows the scatter plot and the QQ plot of the fair market values predicted by GB2 regression and those calculated by Monte Carlo. The plots show that some values predicted by GB2 regression are large. Fitting at the right tail is off. The results indicate that representative policies selected by hierarchical k-means are not good for GB2 regression models.

9.4 Summary

In this chapter, we introduced and implemented the GB2 distribution to model the fair market values in order to capture the skewness. With four parameters, the GB2 distribution is flexible enough to accommodate different forms of skewness of the data. This is the primary advantage of the GB2 regression model over the ordinary and the universal kriging methods. For more information about GB2 regression models, readers are referred to Gan and Valdez (2018c), Cummins et al. (1990), and Sun et al. (2008).

Table 9.1 summarizes the validation measures produced by the GB2 regression model with different experimental design methods. From the table, we see that the GB2 regression model works well with Latin hypercube sampling.

TABLE 9.1: Validation measures of GB2 regression with different experimental design methods.

	LHS	cLHS	Hkmeans
PE	−3.16%	10.11%	9.91%
R^2	0.7986	0.6595	0.4917

10

Rank Order Kriging

In Chapters 7, 8, and 9, we introduced kriging methods and a GB2 regression model for modeling fair market values. Kriging methods do not fit well skewed data but it depends only on a few parameters that can be estimated straightforwardly. GB2 regression can handle skewed data but its parameter estimation can be quite challenging. In this chapter, we introduce the rank order kriging method, which can handle skewed data and depends only on a few parameters.

10.1 Description of the Model

In this section, we describe the rank order kriging method. In the rank order kriging method, standardized ranks of fair market values are estimated and back transformed to the original scale. Additional details of the procedure of this method are shown in Figure 10.1.

Let $\mathbf{z}_1$, $\mathbf{z}_2$, ..., $\mathbf{z}_k$ be the k representative policies and let v_1, v_2, ..., v_k be the corresponding fair market values. For $j = 1, 2, \ldots, k$, let u_j be the standardized rank of v_j, i.e.,

$$u_j = u(\mathbf{z}_j) = \frac{r(v_j)}{k}, \tag{10.1}$$

where $r(v_j) \in \{1, 2, \ldots, k\}$ is the rank order of v_j, that is, $r(v_j)$ is the position of v_j when $v_1, v_2, \ldots, v_k$ are arranged in ascending order. In particular, if v_j is the minimum, then $r(v_j) = 1$ and if v_j is the maximum, then $r(v_j) = k$.

Let $\mathbf{x}_1, \mathbf{x}_2, \ldots, \mathbf{x}_n$ be the variable annuity policies in a portfolio, where n is the number of policies in the portfolio. Under an ordinary kriging model, the standardized rank of the fair market value of the guarantees embedded in the ith policy $\mathbf{x}_i$ is estimated as (Cressie, 1993):

$$\widehat{u}(\mathbf{x}_i) = \sum_{j=1}^{m} w_{ij} u(\mathbf{z}_j), \tag{10.2}$$

where $u(\mathbf{z}_j)$ is the standardized rank order of the jth representative policy as defined in Equation (10.1) and $w_{i1}, w_{i2}, \cdots, w_{im}$ are the kriging weights.

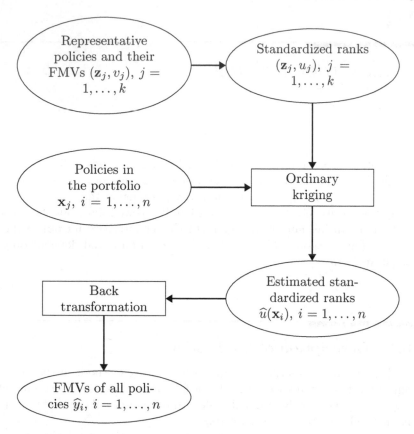

FIGURE 10.1: A sketch of the rank order kriging method.

As shown in Chapter 7, the kriging weights are obtained by solving the following linear equation system:

$$
\begin{pmatrix}
V_{11} & \cdots & V_{1m} & 1 \\
\vdots & \ddots & \vdots & \vdots \\
V_{m1} & \cdots & V_{mm} & 1 \\
1 & \cdots & 1 & 0
\end{pmatrix}
\cdot
\begin{pmatrix}
w_{i1} \\
\vdots \\
w_{im} \\
\theta_i
\end{pmatrix}
=
\begin{pmatrix}
D_{i1} \\
\vdots \\
D_{im} \\
1
\end{pmatrix},
\tag{10.3}
$$

where θ_i is the Lagrange multiplier to ensure the sum of the kriging weights equal to one. In the above equation, V_{ls} and D_{il} are variograms that describe the degree of spatial dependence of the fair market values. Mathematically, V_{ls} and D_{il} are calculated as

$$
V_{ls} = \gamma(\|\mathbf{z}_l - \mathbf{z}_s\|), \tag{10.4a}
$$
$$
D_{il} = \gamma(\|\mathbf{x}_i - \mathbf{z}_l\|), \tag{10.4b}
$$

where $\|\cdot\|$ is the L^2 norm (i.e., the Euclidean distance) and $\gamma(h)$ is a variogram function. In Chapter 7, we introduced three variogram models: exponential, spherical, and Gaussian. Readers are referred to Chapter 7 for the details of these variogram models.

Once we obtain the estimated ranks of the fair market values, we need to transform them back to the original scales. Before doing that, we need to correct the smoothing effect of these estimates caused by kriging. Yamamoto (2005) proposed a procedure to correct the smoothing effect, but this procedure involves cross-validation, which is time consuming. In our method, we modify the rank order kriging estimates as follows:

$$u^*(\mathbf{x}_i) = \frac{r(\widehat{u}(\mathbf{x}_i))}{n}, \tag{10.5}$$

where $r(\widehat{u}(\mathbf{x}_i))$ is the position of $\widehat{u}(\mathbf{x}_i)$ when $\widehat{u}(\mathbf{x}_1)$, $\widehat{u}(\mathbf{x}_2)$, ..., $\widehat{u}(\mathbf{x}_n)$ are arranged in ascending order. The modification is necessary because some ranks estimated by ordinary kriging are out of the interval $(0, 1]$.

The modified rank order kriging estimates are transformed back to the original scale. This can be done by an interpolation method. In particular, we use the following back-transformation:

$$y_i^* = v_{i_1} + \frac{u^*(\mathbf{x}_i) - u_{i_1}}{u_{i_2} - u_{i_1}}(v_{i_2} - v_{i_1}), \tag{10.6}$$

where i_1 and i_2 are indices such that u_{i_1} and u_{i_2} are the ranks closest to $u^*(\mathbf{x}_i)$. The estimated fair market values y_1^*, y_2^*, ..., y_n^* usually contain biases due to the data transformation and are therefore adjusted as follows:

$$\widehat{y}_i = \widehat{c}y_i^*, \tag{10.7}$$

where $\widehat{c}$ is the ratio of an estimate of the mean of the fair market values over the mean of the fair market values obtained from the kriging model, i.e.,

$$\widehat{c} = \frac{\frac{1}{k}\sum_{i=1}^{k} v_i}{\frac{1}{n}\sum_{i=1}^{n} y_i^*}.$$

10.2 Implementation

In this section, we implement the rank order kriging method. We will use the ordinary kriging functions from Chapter 7.

The rank order kriging method is implemented as follows:

```
1  rokrig <- function(Z, y, X, varmodel) {
2    # Perform rank ordinary kriging prediction
3    #
```

```
 4    # Args:
 5    #   Z: a kxd matrix
 6    #   y: a vector of length k
 7    #   X: a nxd matrix
 8    #   varmodel: a variogram model
 9    #
10    # Returns:
11    #   a vector of predicted values for X
12
13    # get standardized rank orders
14    u <- rank(y) / length(y)
15
16    # perform ordinary kriging
17    uhat <- okrig(Z, u, X, varmodel)
18
19    # back transformation
20    require(Hmisc)
21    uhat2 <- rank(uhat) / length(uhat)
22    yhat <- approxExtrap(u, y, rule=2, xout=uhat2)
23    mu <- mean(y)
24    mue <- mean(yhat$y)
25    yhat <- yhat$y * mu / mue
26
27    return(yhat)
28 }
```

In the above implementation, we followed the rank order kriging method sketched in Figure 10.1. We used the R function `rank` to obtain the rank orders of the input data. For back-transformation, we used the function `approxExtrap` from the R package `Hmisc` to do the interpolation as specified in Equation (10.6). The ordinary kriging function `okrig` was implemented in Chapter 7.

10.3   Applications

In this section, we apply the rank order kriging method to predict fair market values of the synthetic portfolio described in Appendix A.

For all the tests, we load and preprocess the data as follows:

```
1 > inforce <- read.csv("inforce.csv")
2 >
3 > vNames <- c("gbAmt", "gmwbBalance", "withdrawal",
        paste("FundValue", 1:10, sep=""))
4 >
5 > age <- with(inforce, (currentDate-birthDate)/365)
```

```
6 > ttm <- with(inforce, (matDate - currentDate)/365)
7 >
8 > datN <- cbind(inforce[,vNames], data.frame(age=age,
    ttm=ttm))
9 > datC <- inforce[,c("gender", "productType")]
10 > dat <- cbind(datN, datC)
11 >
12 > greek <- read.csv("Greek.csv")
13 > greek <- greek[order(greek$recordID),]
14 >
15 > # prepare data
16 > X <- model.matrix( ~ ., data=dat)[,-1]
17 > colnames(X)
18  [1] "gbAmt"             "gmwbBalance"       "withdrawal"
19  [4] "FundValue1"        "FundValue2"        "FundValue3"
20  [7] "FundValue4"        "FundValue5"        "FundValue6"
21 [10] "FundValue7"        "FundValue8"        "FundValue9"
22 [13] "FundValue10"       "age"               "ttm"
23 [16] "genderM"           "productTypeABRU"   "
    productTypeABSU"
24 [19] "productTypeDBAB"   "productTypeDBIB"   "
    productTypeDBMB"
25 [22] "productTypeDBRP"   "productTypeDBRU"   "
    productTypeDBSU"
26 [25] "productTypoDBWB"   "productTypeIBRP"   "
    productTypeIBRU"
27 [28] "productTypeIBSU"   "productTypeMBRP"   "
    productTypeMBRU"
28 [31] "productTypeMBSU"   "productTypeWBRP"   "
    productTypeWBRU"
29 [34] "productTypeWBSU"
30 >
31 > vMin <- apply(X, 2, min)
32 > vMax <- apply(X, 2, max)
33 > X <- (X - matrix(vMin, nrow=nrow(X), ncol= ncol(X),
    byrow=TRUE)) / matrix(vMax-vMin, nrow=nrow(X), ncol=
    ncol(X), byrow=TRUE)
34 > summary(X[,1:10])
35      gbAmt            gmwbBalance         withdrawal
36  Min.   :0.0000   Min.   :0.00000   Min.   :0.00000
37  1st Qu.:0.1382   1st Qu.:0.00000   1st Qu.:0.00000
38  Median :0.2699   Median :0.00000   Median :0.00000
39  Mean   :0.2806   Mean   :0.07232   Mean   :0.04389
40  3rd Qu.:0.4020   3rd Qu.:0.00000   3rd Qu.:0.00000
41  Max.   :1.0000   Max.   :1.00000   Max.   :1.00000
42   FundValue1         FundValue2         FundValue3
43  Min.   :0.000000  Min.   :0.000000  Min.   :0.000000
44  1st Qu.:0.000000  1st Qu.:0.000000  1st Qu.:0.000000
45  Median :0.009006  Median :0.009942  Median :0.008509
```

```
46  Mean     :0.028877   Mean     :0.030847   Mean     :0.029946
47  3rd Qu.:0.042547     3rd Qu.:0.045555     3rd Qu.:0.041759
48  Max.     :1.000000   Max.     :1.000000   Max.     :1.000000
49     FundValue4           FundValue5           FundValue6
50  Min.     :0.000000   Min.     :0.00000    Min.     :0.000000
51  1st Qu.:0.000000     1st Qu.:0.00000      1st Qu.:0.000000
52  Median :0.008731     Median :0.01466      Median :0.009804
53  Mean     :0.029978   Mean     :0.04256    Mean     :0.030445
54  3rd Qu.:0.042890     3rd Qu.:0.06495      3rd Qu.:0.044965
55  Max.     :1.000000   Max.     :1.00000    Max.     :1.000000
56     FundValue7
57  Min.     :0.00000
58  1st Qu.:0.00000
59  Median :0.01040
60  Mean     :0.03388
61  3rd Qu.:0.04897
62  Max.     :1.00000
```

Here we normalized the matrix X to prevent variables with large magnitudes
from dominating the distances.

10.3.1   Rank Order Kriging with Latin Hypercube Sampling

In this subsection, we test the performance of rank order kriging with representative policies selected by Latin hypercube sampling.

The following output shows the performance of rank order kriging with an exponential variogram:

```
 1 > S <- read.table("lhs.csv", sep=",")
 2 > S <- S[,2]
 3 > Z <- X[S,]
 4 > y <- greek$fmv[S]/1000
 5 >
 6 > # fit variogram
 7 > u <- rank(y) / length(y)
 8 > {t1 <- proc.time()
 9 + res <- fitVarModel(Z, u, expVM, 100)
10 + proc.time()-t1}
11    user   system elapsed
12    1.64     0.09     1.73
13 > res
14           a            b            c
15  8.22190251  -0.01924887   0.21166846
16 >
17 > {t1 <- proc.time()
18 + yhat <- rokrig(Z, y, X, function(h) {(expVM(h, res[1],
       res[2], res[3]))})
19 + proc.time()-t1}
```

```
20      user    system  elapsed
21   220.77    12.31   233.86
22 > calMeasure(greek$fmv/1000, yhat)
23 $pe
24 [1]  0.05920789
25
26 $r2
27 [1]  0.6277359
```

From the output, we see that the runtime is similar to that of the ordinary kriging method. The validation measures show that the results produced by rank order kriging are better than those produced by ordinary kriging with the same settings. Figure 10.2 shows the empirical variogram and the fitted exponential variogram based on the standardized rank orders. From this figure, we see that the empirical variogram based on standardized rank orders has an overall stable structure.

Fit variogram model

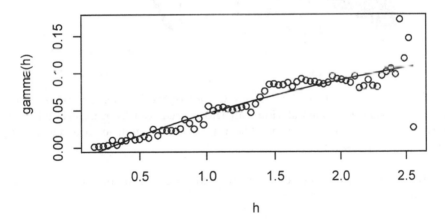

FIGURE 10.2: The empirical variogram and the fitted exponential variogram based on standardized rank orders.

The scatter plot and the QQ plot shown in Figure 10.3 were produced by the following code:

```
1 > png("explhs.png", width=8, height=4, units="in", res
     =300)
2 > par(mfrow=c(1,2), mar=c(4,4,1,1))
3 > plot(greek$fmv/1000, yhat, xlab="FMV(MC)", ylab="FMV(
     ROK)")
```

```
4 > abline(0,1)
5 > qqplot(greek$fmv/1000, yhat, xlab="FMV(MC)", ylab="FMV
     (ROK)")
6 > abline(0,1)
7 > dev.off()
8 null device
9            1
```

The scatter plot shows that some rank orders predicted by ordinary kriging
are different from corresponding values obtained from Monte Carlo.

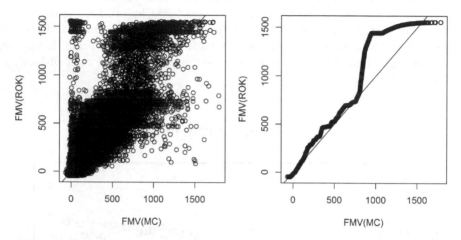

FIGURE 10.3: A scatter plot and a QQ plot of the fair market values calculated by Monte Carlo simulation and those predicted by rank order kriging
with an exponential variogram. The representative policies are selected by
Latin hypercube sampling.

To see the performance of rank order kriging with a spherical variogram,
we change the code as follows:

```
1 > u <- rank(y) / length(y)
2 > {t1 <- proc.time()
3 + res <- fitVarModel(Z, u, sphVM, 100)
4 + proc.time()-t1}
5    user   system elapsed
6    0.44     0.03    0.47
7 > res
8            a          b          c
9  2.86346256 -0.01700845  0.12382655
10 >
11 > {t1 <- proc.time()
12 + yhat <- rokrig(Z, y, X, function(h) {(sphVM(h, res[1],
     res[2], res[3]))})
13 + proc.time()-t1}
```

```
14    user   system elapsed
15  221.20    12.39  234.03
16 > calMeasure(greek$fmv/1000, yhat)
17 $pe
18 [1]  0.05920789
19
20 $r2
21 [1]  -1.075108
22
23 >
24 > png("sphlhs.png", width=8, height=4, units="in", res
       =300)
25 > par(mfrow=c(1,2), mar=c(4,4,1,1))
26 > plot(greek$fmv/1000, yhat, xlab="FMV(MC)", ylab="FMV(
       ROK)")
27 > abline(0,1)
28 > qqplot(greek$fmv/1000, yhat, xlab="FMV(MC)", ylab="FMV
       (ROK)")
29 > abline(0,1)
30 > dev.off()
31 null device
32             1
```

The R^2 in the output shows that the model does not fit the data well. This is confirmed by the scatter plot shown in Figure 10.4.

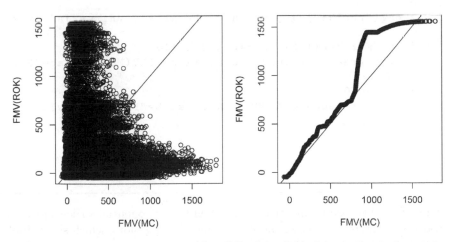

FIGURE 10.4: A scatter plot and a QQ plot of the fair market values calculated by Monte Carlo simulation and those predicted by rank order kriging with a spherical variogram. The representative policies are selected by Latin hypercube sampling.

The following output shows the performance of rank order kriging with a
Gaussian variogram:

```
 1  > u <- rank(y) / length(y)
 2  > {t1 <- proc.time()
 3  + res <- fitVarModel(Z, u, gauVM, 100)
 4  + proc.time()-t1}
 5     user  system elapsed
 6     0.53    0.42    0.95
 7  > res
 8               a              b              c
 9   2.3931670078 -0.0001908393  0.1061957985
10  >
11  > {t1 <- proc.time()
12  + yhat <- rokrig(Z, y, X, function(h) {(gauVM(h, res[1],
        res[2], res[3]))})
13  + proc.time()-t1}
14     user  system elapsed
15   223.29   13.54  238.86
16  > calMeasure(greek$fmv/1000, yhat)
17  $pe
18  [1] 0.05920789
19
20  $r2
21  [1] 0.6257439
22
23  >
24  > png("gaulhs.png", width=8, height=4, units="in", res
        =300)
25  > par(mfrow=c(1,2), mar=c(4,4,1,1))
26  > plot(greek$fmv/1000, yhat, xlab="FMV(MC)", ylab="FMV(
        ROK)")
27  > abline(0,1)
28  > qqplot(greek$fmv/1000, yhat, xlab="FMV(MC)", ylab="FMV
        (ROK)")
29  > abline(0,1)
30  > dev.off()
31  null device
32             1
```

The R^2 in this case is similar to that when the exponential variogram model
was used. Figure 10.5 shows the scatter and the QQ plots, which show similar
patterns as in the case when the exponential variogram model was used.

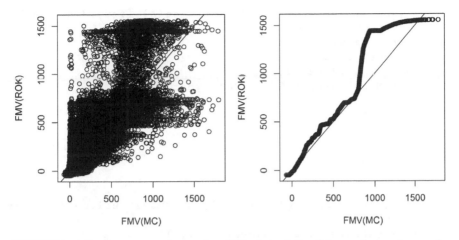

FIGURE 10.5: A scatter plot and a QQ plot of the fair market values calculated by Monte Carlo simulation and those predicted by rank order kriging with a Gaussian variogram. The representative policies are selected by Latin hypercube sampling.

10.3.2 Rank Order Kriging with Conditional Latin Hypercube Sampling

In this subsection, we present results when the rank order kriging method is fitted to representative policies selected by conditional Latin hypercube sampling.

The following output shows the results when the exponential variogram model was used:

```
 1 > S <- read.table("clhs.csv", sep=",")
 2 > S <- S[,2]
 3 > Z <- X[S,]
 4 > y <- greek$fmv[S]/1000
 5 >
 6 > # fit variogram
 7 > u <- rank(y) / length(y)
 8 > {t1 <- proc.time()
 9 + res <- fitVarModel(Z, u, expVM, 100)
10 + proc.time()-t1}
11    user   system  elapsed
12    0.36     0.08     0.45
13 > res
14            a             b             c
15  3.42180549  -0.01874628   0.12736294
16 >
17 > {t1 <- proc.time()
```

```
18 + yhat <- rokrig(Z, y, X, function(h) {(expVM(h, res[1],
      res[2], res[3]))})
19 + proc.time()-t1}
20    user  system elapsed
21  240.60    13.62   267.82
22 > calMeasure(greek$fmv/1000, yhat)
23 $pe
24 [1] -0.01949413
25
26 $r2
27 [1] 0.1832477
28
29 >
30 > png("expclhs.png", width=8, height=4, units="in", res
      =300)
31 > par(mfrow=c(1,2), mar=c(4,4,1,1))
32 > plot(greek$fmv/1000, yhat, xlab="FMV(MC)", ylab="FMV(
      ROK)")
33 > abline(0,1)
34 > qqplot(greek$fmv/1000, yhat, xlab="FMV(MC)", ylab="FMV
      (ROK)")
35 > abline(0,1)
36 > dev.off()
37 null device
38             1
```

From the output, we see that the R^2 is low. The scatter and QQ plots are
shown in Figure 10.6. From the QQ plot, we see that the model does not fit
the right tail well.

The following output shows the performance of rank order kriging with a
spherical variogram:

```
1 > u <- rank(y) / length(y)
2 > {t1 <- proc.time()
3 + res <- fitVarModel(Z, u, sphVM, 100)
4 + proc.time()-t1}
5    user  system elapsed
6    0.40    0.08    0.48
7 > res
8          a          b          c
9  1.9781221 -0.0126125  0.0984095
10 >
11 > {t1 <- proc.time()
12 + yhat <- rokrig(Z, y, X, function(h) {(sphVM(h, res[1],
      res[2], res[3]))})
13 + proc.time()-t1}
14    user  system elapsed
15  232.66    13.34   249.10
16 > calMeasure(greek$fmv/1000, yhat)
```

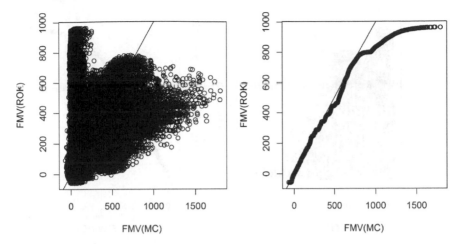

FIGURE 10.6: A scatter plot and a QQ plot of the fair market values calculated by Monte Carlo simulation and those predicted by rank order kriging with an exponential variogram. The representative policies are selected by conditional Latin hypercube sampling.

```
17 $pe
18 [1]  -0.01949413
19
20 $r2
21 [1]  -0.733061
22
23 >
24 > png("sphclhs.png", width=8, height=4, units="in", res
      =300)
25 > par(mfrow=c(1,2), mar=c(4,4,1,1))
26 > plot(greek$fmv/1000, yhat, xlab="FMV(MC)", ylab="FMV(
      ROK)")
27 > abline(0,1)
28 > qqplot(greek$fmv/1000, yhat, xlab="FMV(MC)", ylab="FMV
      (ROK)")
29 > abline(0,1)
30 > dev.off()
31 null device
32           1
```

In this case, the R^2 is negative, indicating that the fit is not good. This is confirmed by the scatter and QQ plots shown in Figure 10.7. The scatter plot shows that the predicted rank orders of many policies are off.

When a Gaussian variogram model was used, the output is shown below:

```
1 > u <- rank(y) / length(y)
2 > {t1 <- proc.time()
```

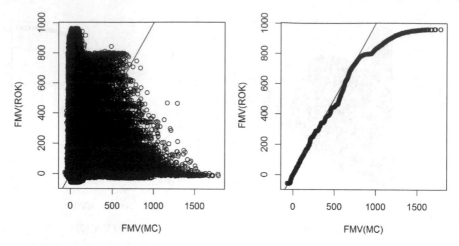

FIGURE 10.7: A scatter plot and a QQ plot of the fair market values calculated by Monte Carlo simulation and those predicted by rank order kriging
with a spherical variogram. The representative policies are selected by conditional Latin hypercube sampling.

```
 3 + res <- fitVarModel(Z, u, gauVM, 100)
 4 + proc.time()-t1}
 5    user  system elapsed
 6    0.43    0.05    0.48
 7 > res
 8              a            b            c
 9 1.766503565 0.002805863 0.085163849
10 >
11 > {t1 <- proc.time()
12 + yhat <- rokrig(Z, y, X, function(h) {(gauVM(h, res[1],
       res[2], res[3]))})
13 + proc.time()-t1}
14    user  system elapsed
15  224.99   12.78  239.74
16 > calMeasure(greek$fmv/1000, yhat)
17 $pe
18 [1]  -0.01949413
19
20 $r2
21 [1]  0.7560632
22
23 >
24 > png("gauclhs.png", width=8, height=4, units="in", res
       =300)
25 > par(mfrow=c(1,2), mar=c(4,4,1,1))
26 > plot(greek$fmv/1000, yhat, xlab="FMV(MC)", ylab="FMV(
       ROK)")
```

```
27 > abline(0,1)
28 > qqplot(greek$fmv/1000, yhat, xlab="FMV(MC)", ylab="FMV
     (ROK)")
29 > abline(0,1)
30 > dev.off()
31 null device
32           1
```

The R^2 in the output shows that the results are better than the previous two cases. The scatter plot given in Figure 10.8 shows that the fit is better although the predicted rank orders of some policies are off.

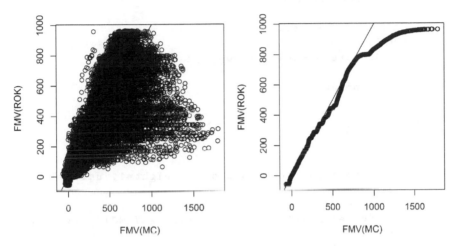

FIGURE 10.8: A scatter plot and a QQ plot of the fair market values calculated by Monte Carlo simulation and those predicted by rank order kriging with a Gaussian variogram. The representative policies are selected by conditional Latin hypercube sampling.

10.3.3 Rank Order Kriging with Hierarchical k-Means

In this subsection, we test how rank order kriging performs when hierarchical k-means is used to select representative policies.

The following output shows the performance of rank order kriging with an exponential variogram:

```
1 > S <- read.table("hkmeans.csv", sep=",")
2 > S <- S[,2]
3 > Z <- X[S,]
4 > y <- greek$fmv[S]/1000
5 >
6 > # fit variogram
7 > u <- rank(y) / length(y)
```

```
 8 > {t1 <- proc.time()
 9 + res <- fitVarModel(Z, u, expVM, 100)
10 + proc.time()-t1}
11    user  system elapsed
12    0.30    0.04    0.34
13 > res
14                a               b               c
15  9.397395353 -0.008279924  0.215081056
16 >
17 > {t1 <- proc.time()
18 + yhat <- rokrig(Z, y, X, function(h) {(expVM(h, res[1],
        res[2], res[3]))})
19 + proc.time()-t1}
20    user  system elapsed
21  229.70   13.82  246.76
22 > calMeasure(greek$fmv/1000, yhat)
23 $pe
24 [1] 0.0555428
25
26 $r2
27 [1] 0.885311
28
29 >
30 > png("expkmeans.png", width=8, height=4, units="in",
        res=300)
31 > par(mfrow=c(1,2), mar=c(4,4,1,1))
32 > plot(greek$fmv/1000, yhat, xlab="FMV(MC)", ylab="FMV(
        ROK)")
33 > abline(0,1)
34 > qqplot(greek$fmv/1000, yhat, xlab="FMV(MC)", ylab="FMV
        (ROK)")
35 > abline(0,1)
36 > dev.off()
37 null device
38              1
```

The R^2 shows that rank order kriging produced a good fit when the representative policies were selected by hierarchical k-means. The scatter and QQ plots are given in Figure 10.9, which shows that most of the rank orders were predicted correctly.

When a spherical variogram was used, the output is given below:

```
1 > u <- rank(y) / length(y)
2 > {t1 <- proc.time()
3 + res <- fitVarModel(Z, u, sphVM, 100)
4 + proc.time()-t1}
5    user  system elapsed
6    0.34    0.03    0.38
7 > res
```

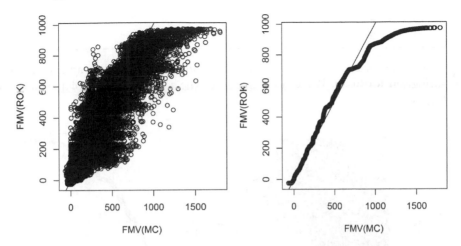

FIGURE 10.9: A scatter plot and a QQ plot of the fair market values calculated by Monte Carlo simulation and those predicted by rank order kriging with an exponential variogram. The representative policies are selected by hierarchical k-means.

```
 8                a             b            c
 9   2.454681472  -0.008816085   0.104432991
10 >
11 > {t1 <- proc.time()
12 + yhat <- rokrig(Z, y, X, function(h) {(sphVM(h, res[1],
       res[2], res[3]))})
13 + proc.time()-t1}
14     user   system  elapsed
15   227.59    13.90   242.98
16 > calMeasure(greek$fmv/1000, yhat)
17 $pe
18 [1] 0.0555428
19
20 $r2
21 [1] 0.7954062
22
23 >
24 > png("sphkmeans.png", width=8, height=4, units="in",
       res=300)
25 > par(mfrow=c(1,2), mar=c(4,4,1,1))
26 > plot(greek$fmv/1000, yhat, xlab="FMV(MC)", ylab="FMV(
       ROK)")
27 > abline(0,1)
28 > qqplot(greek$fmv/1000, yhat, xlab="FMV(MC)", ylab="FMV
       (ROK)")
29 > abline(0,1)
30 > dev.off()
```

```
31 windows
32          2
```

The R^2 shows that the results are not as good as the case when an exponential variogram was used. We also see this from the scatter and QQ plots shown in Figure 10.10.

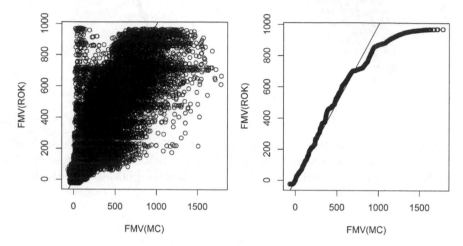

FIGURE 10.10: A scatter plot and a QQ plot of the fair market values calculated by Monte Carlo simulation and those predicted by rank order kriging with a spherical variogram. The representative policies are selected by hierarchical k-means.

The following output shows the performance of rank order kriging with a Gaussian variogram:

```
 1 > u <- rank(y) / length(y)
 2 > {t1 <- proc.time()
 3 + res <- fitVarModel(Z, u, gauVM, 100)
 4 + proc.time()-t1}
 5    user   system elapsed
 6    0.36     0.06    0.44
 7 > res
 8               a             b             c
 9 2.184985569 0.007982675 0.090078396
10 >
11 > {t1 <- proc.time()
12 + yhat <- rokrig(Z, y, X, function(h) {(gauVM(h, res[1],
        res[2], res[3]))})
13 + proc.time()-t1}
14    user   system elapsed
15  226.03    13.27  241.31
16 > calMeasure(greek$fmv/1000, yhat)
17 $pe
```

```
18 [1]  0.0555428
19
20 $r2
21 [1]  0.8525405
22
23 >
24 > png("gaukmeans.png", width=8, height=4, units="in",
      res=300)
25 > par(mfrow=c(1,2), mar=c(4,4,1,1))
26 > plot(greek$fmv/1000, yhat, xlab="FMV(MC)", ylab="FMV(
      ROK)")
27 > abline(0,1)
28 > qqplot(greek$fmv/1000, yhat, xlab="FMV(MC)", ylab="FMV
      (ROK)")
29 > abline(0,1)
30 > dev.off()
31 windows
32          2
```

The R^2 shows that the results are better than the previous case when a spherical variogram was used. This can also be seen from the scatter and the QQ plots given in Figure 10.11.

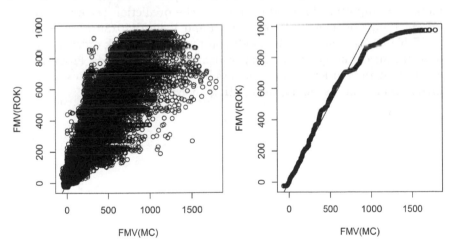

FIGURE 10.11: A scatter plot and a QQ plot of the fair market values calculated by Monte Carlo simulation and those predicted by rank order kriging with a Gaussian variogram. The representative policies are selected by hierarchical k-means.

TABLE 10.1: Validation measures of rank order kriging with different combinations of variogram models and experimental design methods.

	PE			R^2		
	LHS	cLHS	Hkmeans	LHS	cLHS	Hkmeans
Exponential	5.92%	−1.95%	5.55%	0.6277	0.1832	0.8853
Spherical	5.92%	−1.95%	5.55%	−1.0751	−0.7331	0.7954
Gaussian	5.92%	−1.95%	5.55%	0.6257	0.7561	0.8525

10.4 Summary

In this chapter, we introduced and implemented the rank order kriging method for predicting fair market values. Rank order kriging consists of three main steps: convert fair market values to standardized rank orders, perform ordinary kriging on these rank orders, and transform the estimated rank orders back to the original scale. While rank order kriging is developed to handle skewed data, back-transformation poses challenges. There are several ways to back transform the estimates and using an appropriate back-transformation method is important in order to improve the prediction accuracy. For more information, readers are referred to Gan and Valdez (2018d).

Table 10.1 summarizes the validation measures of rank order kriging with different variogram models and experimental design methods. From the table, we see that rank order kriging works well with an exponential variogram and hierarchical k-means among the existing choices.

11

Linear Model with Interactions

In this chapter, we introduce linear regression models to predict the fair market values. Linear regression models are simple models when compared to the models introduced in previous chapters. However, we add many interaction terms to improve the prediction accuracy. With so many possible interaction terms, we employ overlapped group-lasso to select the important interaction terms.

11.1 Description of the Model

In this section, we introduce a first-order interaction model, which is an extension of the multiple linear regression model by adding some interaction terms. Let Y be a continuous response variable and let X_1, X_2, ..., X_p be p explanatory variables. Then the first-order interaction model is given by (Lim and Hastie, 2015):

$$E[Y|X_1, X_2, \ldots, X_p] = \beta_0 + \sum_{j=1}^{p} \beta_j X_j + \sum_{s<t} \beta_{s:t} X_{s:t}, \qquad (11.1)$$

where

- $X_{s:t} = X_s X_t$ denotes the interaction effect between X_s and X_t, and

- X_1, X_2, ..., X_p denote the main effects.

Main effects model deviations from the global mean and interaction effects model deviations from the main effects.

An interaction model is said to be a hierarchical interaction model if it satisfies one of the following conditions:

- an interaction can exist only if both its main effects are present (strong hierarchy), or

- an interaction can exist as long as either of its main effects is present (weak hierarchy).

Since it rarely makes sense to have interactions without main effects, hierarchical interaction models are usually preferred.

The first-order interaction model given in Equation (11.1) has a simple form. However, learning interactions is quite challenging, especially when there are many explanatory variables. For example, the number of pairwise interaction terms among 34 variables is $34 \times 33/2 = 561$, which may exceed the number of training samples. Including all the pairwise interaction terms in the model may overfit the data. To avoid overfitting the data, we need to select important interactions only.

The overlapped group-lasso proposed by Lim and Hastie (2015) can be used to select interactions and produce a hierarchical interaction model automatically. In what follows, we describe the overlapped group-lasso method.

The overlapped group-lasso extends the group-lasso (Yuan and Lin, 2006) by adding an overlapped group-lasso penalty. The group-lasso is a general version of the popular lasso (least absolute shrinkage and selection operator) proposed by Tibshirani (1996).

Let $\mathbf{y} = (y_1, y_2, \ldots, y_n)'$ denote the vector of responses and let $\mathbf{X}$ denote the design matrix. Then the lasso is formulated as the following optimization problem:

$$\widehat{\boldsymbol{\beta}}^{LASSO} = \arg\min_{\boldsymbol{\beta}} \left(\frac{1}{2} \|\mathbf{y} - \mathbf{X}\boldsymbol{\beta}\|_2^2 + \lambda\|\boldsymbol{\beta}\|_1 \right), \qquad (11.2)$$

where $\|\cdot\|_2$ denotes the ℓ^2-norm, $\|\cdot\|_1$ denotes the ℓ^1-norm, and λ is a tuning parameter that controls the amount of regularization. The ℓ^1-norm has a nice property that it induces sparsity in the solution by forcing some coefficients to zero. A larger value of λ implies more regularization, i.e., more coefficients will be zero.

The lasso is designed for selecting individual explanatory variables but not for factors. The group-lasso was proposed to select important factors. Suppose that there are p groups of explanatory variables. For $j = 1, 2, \ldots, p$, let $\mathbf{X}_j$ denote the feature matrix for group j. The group-lasso is formulated as follows:

$$\widehat{\boldsymbol{\beta}}^{GLASSO} = \arg\min_{\boldsymbol{\beta}} \left(\frac{1}{2} \|\mathbf{y} - \beta_0 \mathbf{1} - \sum_{j=1}^{p} \mathbf{X}_j \boldsymbol{\beta}_j\|_2^2 + \lambda \sum_{j=1}^{p} \gamma_j \|\boldsymbol{\beta}_j\|_2 \right), \quad (11.3)$$

where $\mathbf{1}$ is a vector of ones and $\lambda, \gamma_1, \ldots, \gamma_p$ are tuning parameters. When each group contains one continuous variable, the group-lasso reduces to the lasso. The group-lasso has an attractive property that if $\widehat{\boldsymbol{\beta}}_j$ is nonzero, then all its components are typically nonzero.

The group-lasso described above may not lead to hierarchical interaction models. In order to obtain hierarchical interaction models, one way is to add an overlapped group-lasso penalty to the loss function. The resulting method is called the overlapped group-lasso, which is formulated as the following constrained optimization problem (Lim and Hastie, 2015):

$$\hat{\boldsymbol{\beta}}^{OGLASSO}$$

$$= \arg\min_{\boldsymbol{\beta}} \left(\frac{1}{2} \left\| \mathbf{y} - \beta_0 \mathbf{1} - \sum_{j=1}^{p} \mathbf{X}_j \boldsymbol{\beta}_j - \sum_{s<t} (\mathbf{X}_s \tilde{\boldsymbol{\beta}}_s + \mathbf{X}_t \tilde{\boldsymbol{\beta}}_t + \mathbf{X}_{s:t} \boldsymbol{\beta}_{s:t}) \right\|_2^2 \right.$$

$$\left. + \lambda \left(\sum_{j=1}^{p} \|\boldsymbol{\beta}_j\|_2 + \sum_{s<t} \sqrt{L_s \|\tilde{\boldsymbol{\beta}}_s\|_2^2 + L_t \|\tilde{\boldsymbol{\beta}}_t\|_2^2 + \|\boldsymbol{\beta}_{s:t}\|_2^2} \right) \right), \qquad (11.4)$$

subject to the following sets of constraints:

- if X_j is categorical,

$$\sum_{l=1}^{m_j} \beta_j^{(l)} = 0, \quad \sum_{l=1}^{m_j} \tilde{\beta}_j^{(l)} = 0,$$

- if X_j is categorical and X_t is continuous,

$$\sum_{l=1}^{m_j} \beta_{t:j}^{(l)} = 0,$$

- if X_j and X_t are both categorical,

$$\sum_{l=1}^{m_j} \beta_{t:j}^{(l,k)} = 0 \quad \forall k, \quad \sum_{k=1}^{m_t} \beta_{t:j}^{(l,k)} = 0 \quad \forall l.$$

Here m_j is the number of levels of X_j, m_t is the number of levels of X_t, $\beta_j^{(l)}$ is the lth entry of $\boldsymbol{\beta}_j$, $\tilde{\beta}_j^{(l)}$ is the lth entry of $\tilde{\boldsymbol{\beta}}_j$, and $\beta_{t:j}^{(l,k)}$ is the lkth entry of $\boldsymbol{\beta}_{t:j}$. In addition, $\mathbf{X}_1$, $\mathbf{X}_2$, ..., $\mathbf{X}_p$ denote the feature matrices of the p group of variables. If X_j is continuous, then $\mathbf{X}_j$ is just a one-column matrix containing the values of X_j. If X_j is categorical, then $\mathbf{X}_j$ contains all the dummy variables associated with X_j. The matrix $\mathbf{X}_{s:t}$ denotes the feature matrix of the interaction term.

11.2 Implementation

To fit an interaction model to the representative policies that satisfy a strong hierarchy, we will use the function `glinternet.cv` from the R package `glinternet`.

The following code loads the package and shows the function:

```
1 > require(glinternet)
2 Loading required package: glinternet
3 Loaded glinternet 1.0.8
4
5 > glinternet.cv
6 function (X, Y, numLevels, nFolds = 10, lambda = NULL,
     nLambda = 50,
7     lambdaMinRatio = 0.01, interactionCandidates = NULL,
         screenLimit = NULL,
8     family = c("gaussian", "binomial"), tol = 1e-05,
         maxIter = 5000,
9     verbose = FALSE, numCores = 1)
```

From the output, we see that the function `glinternet.cv` has a number of arguments. For our purpose, we only need to consider the first three arguments:

1. `X`, which is the data matrix,
2. `Y`, which is the response vector, and
3. `numLevels`, which specifies the number of levels of each explanatory variables.

Note that categorical variables should be represented by consecutive integers starting from zero.

11.3   Applications

In this section, we apply linear models with interactions to predict the fair market values of the synthetic portfolio described in Appendix A.

Before all the tests, we load and prepare the data as follows:

```
1 > inforce <- read.csv("inforce.csv")
2 >
3 > vNames <- c("gbAmt", "gmwbBalance", "withdrawal",
     paste("FundValue", 1:10, sep=""))
4 >
5 > age <- with(inforce, (currentDate-birthDate)/365)
6 > ttm <- with(inforce, (matDate - currentDate)/365)
7 >
8 > datN <- cbind(inforce[,vNames], data.frame(age=age,
     ttm=ttm))
9 > datC <- inforce[,c("gender", "productType")]
10 >
11 > dat <- cbind(datN, gender=as.numeric(datC$gender)-1,
```

```
12 +      productType=as.numeric(datC$productType)-1)
13 > summary(dat)
14       gbAmt           gmwbBalance           withdrawal
15  Min.   : 50002   Min.    :      0   Min.    :      0
16  1st Qu.:179759   1st Qu.:      0   1st Qu.:      0
17  Median :303525   Median :      0   Median :      0
18  Mean   :313507   Mean   : 36141   Mean   : 21928
19  3rd Qu.:427544   3rd Qu.:      0   3rd Qu.:      0
20  Max.   :989205   Max.   :499709   Max.   :499586
21     FundValue1         FundValue2         FundValue3
22  Min.   :      0   Min.   :      0   Min.   :      0
23  1st Qu.:      0   1st Qu.:      0   1st Qu.:      0
24  Median :   8299   Median :   8394   Median :   4942
25  Mean   :  26611   Mean   :  26045   Mean   :  17391
26  3rd Qu.:  39209   3rd Qu.:  38463   3rd Qu.:  24251
27  Max.   :921549   Max.   :844323   Max.   :580753
28     FundValue4         FundValue5         FundValue6
29  Min.   :      0   Min.   :      0   Min.   :      0
30  1st Qu.:      0   1st Qu.:      0   1st Qu.:      0
31  Median :   4225   Median :   7248   Median :   8556
32  Mean   :  14507   Mean   :  21041   Mean   :  26570
33  3rd Qu.:  20756   3rd Qu.:  32112   3rd Qu.:  39241
34  Max.   :483937   Max.   :494382   Max.   :872707
35     FundValue7         FundValue8         FundValue9
36  Min.   :      0   Min.   :      0   Min.   :      0
37  1st Qu.:      0   1st Qu.:      0   1st Qu.:      0
38  Median :   6602   Median :   6255   Median :   5943
39  Mean   :  21506   Mean   :  19990   Mean   :  19647
40  3rd Qu.:  31088   3rd Qu.:  29404   3rd Qu.:  28100
41  Max.   :634819   Max.   :562485   Max.   :663196
42   FundValue10            age                ttm
43  Min.   :      0   Min.   :34.52   Min.    : 0.5863
44  1st Qu.:      0   1st Qu.:42.03   1st Qu.:10.3425
45  Median :   6738   Median :49.45   Median :14.5123
46  Mean   :  21003   Mean   :49.49   Mean   :14.5362
47  3rd Qu.:  31256   3rd Qu.:56.96   3rd Qu.:18.7616
48  Max.   :599675   Max.   :64.46   Max.   :28.5205
49       gender        productType
50  Min.   :0.0   Min.   : 0
51  1st Qu.:0.0   1st Qu.: 4
52  Median :1.0   Median : 9
53  Mean   :0.6   Mean   : 9
54  3rd Qu.:1.0   3rd Qu.:14
55  Max.   :1.0   Max.   :18
56 >
57 > greek <- read.csv("Greek.csv")
58 > greek <- greek[order(greek$recordID),]
```

From the above output, we see that the categorical values are represented by integers starting from zero.

11.3.1   Linear Model with Latin Hypercube Sampling

In this subsection, we test the performance of linear models with interactions based on representative policies selected by Latin hypercube sampling. To do that, we proceed as follows:

```
 1 > S <- read.table("lhs.csv", sep=",")
 2 > S <- S[,2]
 3 >
 4 > y <- greek$fmv[S]/1000
 5 >
 6 > {t1 <- proc.time()
 7 + set.seed(1)
 8 + numLevels <- c(rep(1, ncol(dat)-2), 2, 19)
 9 + fit <- glinternet.cv(dat[S,], y, numLevels)
10 + yhat <- predict(fit, dat)
11 + proc.time()-t1}
12    user   system  elapsed
13   36.36    0.16    36.84
14 > calMeasure(greek$fmv/1000, yhat)
15 $pe
16 [1] -0.5280413
17
18 $r2
19 [1] 0.7685298
20
21 >
22 > png("lassolhs.png", width=8, height=4, units="in", res
        =300)
23 > par(mfrow=c(1,2), mar=c(4,4,1,1))
24 > plot(greek$fmv/1000, yhat, xlab="FMV(MC)", ylab="FMV(
        OK)")
25 > abline(0,1)
26 > qqplot(greek$fmv/1000, yhat, xlab="FMV(MC)", ylab="FMV
        (OK)")
27 > abline(0,1)
28 > dev.off()
29 null device
30            1
```

The percentage error in the output shows that the model underestimated the fair market value of the portfolio. The R^2 indicates the fit is relatively good. Figure 11.1 shows the scatter plot and the QQ plot. The QQ plot shows that the model does not fit the left tail well and the predictions are biased.

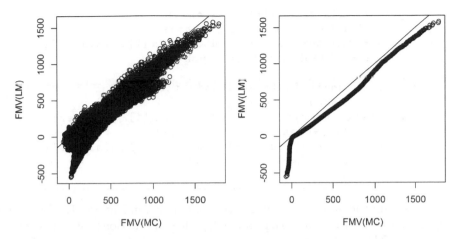

FIGURE 11.1: A scatter plot and a QQ plot of the fair market values calculated by Monte Carlo simulation and those predicted by a linear model with interactions. The representative policies are selected by Latin hypercube sampling.

To see what interaction terms are selected by the algorithm, we can visualize the interaction terms by running the following code:

```
1 coeffs <- coef(fit)
2 vNames = c(names(dat)[16:17], names(dat)[1:15])
3 par(mar=c(7,7,1,1))
4 plot(0,type="n",xlim=c(0, 17), ylim=c(0, 17), axes=FALSE
      ,ann=FALSE)
5 grid(17,17, col="black")
6 vLim <- par("usr")
7 dLen <- (vLim[2] - vLim[1]) / 17
8 box()
9 axis(1, at=seq(vLim[1] + 0.5*dLen, vLim[2], dLen),
      labels=vNames, las=2)
10 axis(2, at=seq(vLim[1] + 0.5*dLen, vLim[2], dLen),
      labels=vNames, las=2)
11
12 mInter <- coeffs$interactions$catcat
13 for(i in 1:nrow(mInter)) {
14   points((mInter[i,1]-0.5)*dLen+vLim[1], (mInter[i
        ,2]-0.5)*dLen+vLim[1], pch=19)
15   points((mInter[i,2]-0.5)*dLen+vLim[1], (mInter[i
        ,1]-0.5)*dLen+vLim[1], pch=19)
16 }
17 mInter <- coeffs$interactions$contcont
18 for(i in 1:nrow(mInter)) {
```

```
19    points((mInter[i,1]+1.5)*dLen+vLim[1], (mInter[i
         ,2]+1.5)*dLen+vLim[1], pch=19)
20    points((mInter[i,2]+1.5)*dLen+vLim[1], (mInter[i
         ,1]+1.5)*dLen+vLim[1], pch=19)
21 }
22 mInter <- coeffs$interactions$catcont
23 for(i in 1:nrow(mInter)) {
24    points((mInter[i,1]-0.5)*dLen+vLim[1], (mInter[i
         ,2]+1.5)*dLen+vLim[1], pch=19)
25    points((mInter[i,2]+1.5)*dLen+vLim[1], (mInter[i
         ,1]-0.5)*dLen+vLim[1], pch=19)
26 }
```

After we run the above code, we see the plot shown in Figure 11.1. From the figure, we see that the variable productType has interactions with many other variables. This makes sense because the variable productType controls how the guarantee payoffs are calculated.

11.3.2 Linear Model with Conditional Latin Hypercube Sampling

To see the performance of linear models with interactions when conditional Latin hypercube sampling is used to select representative policies, we proceed as follows:

```
 1 > S <- read.table("clhs.csv", sep=",")
 2 > S <- S[,2]
 3 >
 4 > y <- greek$fmv[S]/1000
 5 >
 6 > {t1 <- proc.time()
 7 + set.seed(1)
 8 + numLevels <- c(rep(1, ncol(dat)-2), 2, 19)
 9 + fit <- glinternet.cv(dat[S,], y, numLevels)
10 + yhat <- predict(fit, dat)
11 + proc.time()-t1}
12    user  system elapsed
13   40.66    0.20   41.08
14 > calMeasure(greek$fmv/1000, yhat)
15 $pe
16 [1] 0.003771857
17
18 $r2
19 [1] 0.970502
20
21 >
22 > png("lassoclhs.png", width=8, height=4, units="in",
        res=300)
```

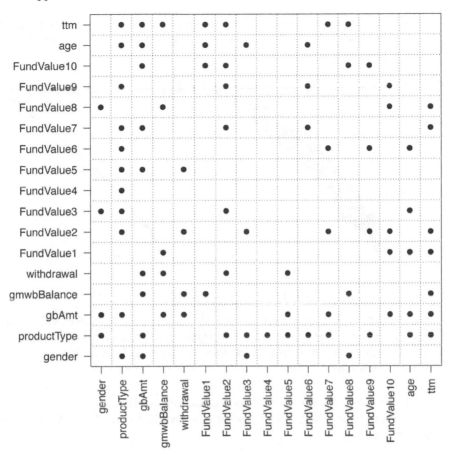

FIGURE 11.2: Pairwise interactions found by the overlapped group-lasso when representative policies were selected by Latin hypercube sampling.

```
23 > par(mfrow=c(1,2), mar=c(4,4,1,1))
24 > plot(greek$fmv/1000, yhat, xlab="FMV(MC)", ylab="FMV(
     OK)")
25 > abline(0,1)
26 > qqplot(greek$fmv/1000, yhat, xlab="FMV(MC)", ylab="FMV
     (OK)")
27 > abline(0,1)
28 > dev.off()
29 windows
30     2
```

The validation measures show that the model produced excellent results. Figure 11.3 shows the scatter plot and the QQ plot of the fair market values.

From the scatter plot, we see that the model did a good job predicting the fair market values. The QQ plot shows that the fit is a little bit off at the right tail.

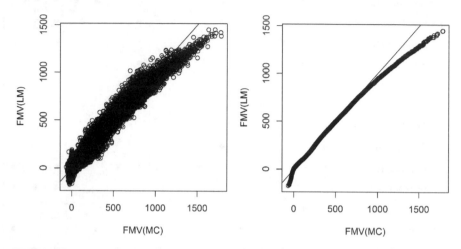

FIGURE 11.3: A scatter plot and a QQ plot of the fair market values calculated by Monte Carlo simulation and those predicted by a linear model with interactions. The representative policies are selected by conditional Latin hypercube sampling.

To visualize the interaction terms, we can use the code given in the previous subsection. The resulting figure is shown in Figure 11.4. We see similar patterns as before.

11.3.3   Linear Model with Hierarchical k-Means

The following output shows the performance of linear models with interactions when hierarchical k-means was used to select representative policies:

```
 1  > S <- read.table("hkmeans.csv", sep=",")
 2  > S <- S[,2]
 3  >
 4  > y <- greek$fmv[S]/1000
 5  >
 6  > {t1 <- proc.time()
 7  + set.seed(1)
 8  + numLevels <- c(rep(1, ncol(dat)-2), 2, 19)
 9  + fit <- glinternet.cv(dat[S,], y, numLevels)
10  + yhat <- predict(fit, dat)
11  + proc.time()-t1}
12     user   system  elapsed
13    30.44     0.02    30.65
14  > calMeasure(greek$fmv/1000, yhat)
```

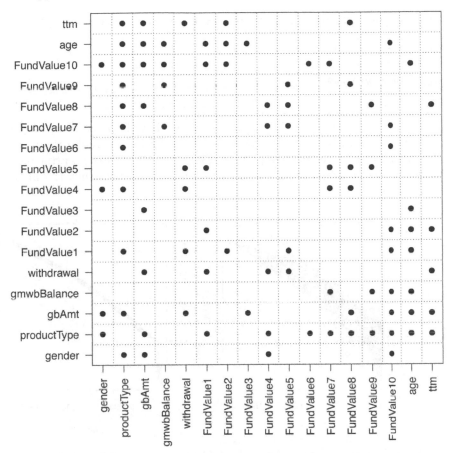

FIGURE 11.4: Pairwise interactions found by the overlapped group-lasso when representative policies were selected by conditional Latin hypercube sampling.

```
15 $pe
16 [1]  0.008927298
17
18 $r2
19 [1]  0.9523106
20
21 >
22 > png("lassohkmeans.png", width=8, height=4, units="in",
       res=300)
23 > par(mfrow=c(1,2), mar=c(4,4,1,1))
24 > plot(greek$fmv/1000, yhat, xlab="FMV(MC)", ylab="FMV(
       LM)")
25 > abline(0,1)
```

```
26 > qqplot(greek$fmv/1000, yhat, xlab="FMV(MC)", ylab="FMV
     (LM)")
27 > abline(0,1)
28 > dev.off()
29 null device
30               1
```

The validation measures in the output show that the results are very good but not as good as the case when conditional Latin hypercube sampling was used to select representative policies. Figure 11.5 shows the scatter plot and the QQ plot corresponding to this case. The plots show that the fit at the tails is off.

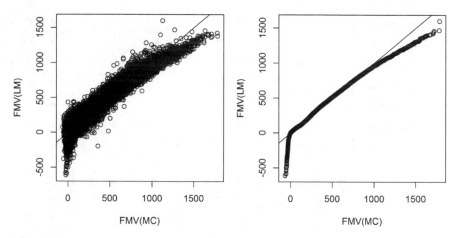

FIGURE 11.5: A scatter plot and a QQ plot of the fair market values calculated by Monte Carlo simulation and those predicted by a linear model with interactions. The representative policies are selected by hierarchical k-means.

Figure 11.6 visualizes the important interaction terms found by the algorithm. From the figure, we see that fewer interaction terms were found as compared to the previous two cases. This indicates that experimental design methods have a big impact on the interaction models.

TABLE 11.1: Validation measures of linear models with different experimental design methods.

	LHS	cLHS	Hkmeans
PE	−52.80%	0.38%	0.89%
R^2	0.7685	0.9705	0.9523

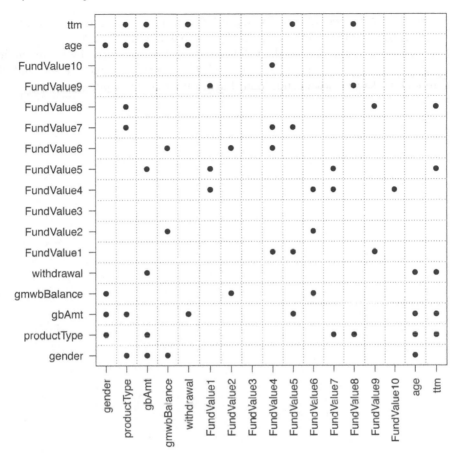

FIGURE 11.6: Pairwise interactions found by the overlapped group-lasso when representative policies were selected by hierarchical k-means.

11.4 Summary

In this chapter, we introduced and demonstrated the use of linear models with interactions to predict fair market values. Our results show that linear models with interactions outperform models described in previous chapters. The results indicate that it is important to model interactions when developing models for predicting fair market values. It is not surprising that this linear model with interaction terms produces excellent results because cash flow projections in Monte Carlo simulation depend on combinations of the variables. See Gan (2018) for the case when random sampling was used to select representative policies. For more information about group-lasso, readers are referred to Lim and Hastie (2015) and Yuan and Lin (2006).

Table 11.1 summarizes the validation measures produced by linear models with different experimental design methods. From the table, we see that the model works well with conditional Latin hypercube sampling.

12

Tree-Based Models

In the previous chapters, we introduced several models for predicting fair market values. In this chapter, we introduce and illustrate the use of tree-based models for this purpose. Tree-based models have increased in popularity as an alternative to traditional regression models. Building such models involves algorithms that repeatedly partition the region of the explanatory variables to create non-overlapping nodes for predictions. Tree-based models have several advantages. For example, they can handle missing data automatically, handle categorical variables naturally, capture nonlinear effects, and handle interactions between variables.

12.1 Description of the Model

Tree-based models divide the predictor space (i.e., the space formed by explanatory variables) into a number of non-overlapping regions and use the mean or the mode of the region as the prediction. As a result, tree-based models can be used to predict continuous or categorical responses. When the response variable is continuous, the resulting tree model is called a regression tree. When the response is categorical, the resulting tree model is called a classification tree. In this section, we give a brief introduction to regression trees. For a detailed treatment, readers are referred to Breiman et al. (1984) and Mitchell (1997).

Let z_1, z_2, ..., z_k denote the representative policies and let y_1, y_2, ..., y_k denote the corresponding fair market values. To build a regression tree, the predictor space is divided into J non-overlapping regions such that the following objective function

$$f(R_1, R_2, \ldots, R_J) = \sum_{j=1}^{J} \sum_{i=1}^{k} I_{R_j}(\mathbf{z}_i)(y_i - \mu_j)^2 \qquad (12.1)$$

is minimized, where I is an indicator function, R_j denotes the set of indices of the observations that belong to the jth box, and μ_j is the mean response of the observations in the jth box.

The idea of the regression tree model is simple. However, there are a large number of ways to divide the predictor space into J boxes when the number of explanatory variable is large. Suppose that there are p explanatory variables and the jth variable has k distinct values. Then there are $p(k-1)$ ways to divide the predictor space into two boxes. Similarly, there are $p(k-2)$ ways to divide the two boxes into three boxes. Hence there are

$$\prod_{j=1}^{J-1}(p(k-j)) = p^{J-1}\prod_{j=1}^{k}(k-j) \tag{12.2}$$

ways to divide the predictor space into J boxes.

To reduce the number of evaluations, the recursive binary splitting method is used to find the approximately optimal division. This method starts with the whole predictor space and repeats using a greedy strategy to select a box and divide it into two until J boxes have been created. At the beginning, this method first splits the whole predictor space into two boxes by finding j and t to minimize the following objective function

$$\sum_{i=1}^{k}\left[I_{R_1(j,t)}(\mathbf{x}_i)(y_i - \mu_1(j,t))^2 + I_{R_2(j,t)}(\mathbf{x}_i)(y_i - \mu_2(j,t))^2\right], \tag{12.3}$$

where

$$R_1(j,t) = \{\mathbf{x} : x_j < t\}, \quad R_2(j,t) = \{\mathbf{x} : x_j > t\},$$

$$\mu_1(j,t) = \frac{1}{|R_1(j,t)|}\sum_{i=1}^{k}I_{R_1(j,t)}(\mathbf{x}_i)y_i,$$

and

$$\mu_2(j,t) = \frac{1}{|R_2(j,t)|}\sum_{i=1}^{k}I_{R_2(j,t)}(\mathbf{x}_i)y_i.$$

The index j ranges from 1 to p and the cutpoint t can take at most $k-1$ distinct values.

The recursive binary splitting method repeats the above process. Suppose that $J-1$ boxes have been created by the binary splitting method. To continue, the method will split one of the $J-1$ boxes into two such that the decrease in the objective function is maximized. The number of objective function evaluations is

$$\sum_{j=1}^{J-1}(p(k-j)) = \frac{(2k-J)(J-1)p}{2},$$

which is not a big number when both k and p are not large.

In terms of predictive accuracy, regression trees generally do not perform to the level of other regression techniques. However, aggregating many regression trees has the potential to improve the predictive accuracy significantly. There are three approaches for aggregating regression trees:

Bagging Bagging is a general-purpose method for reducing the variance of a predictive model. This method works with regression trees as follows. First, we generate M different bootstrapped training datasets, each of which is a random sample. Second, we build a regression tree for each of the M bootstrapped training datasets. Finally, we average all the predictions from the M regression trees:

$$f_{bag}(\mathbf{x}) = \frac{1}{M} \sum_{l=1}^{M} f_l(\mathbf{x}), \qquad (12.4)$$

where $f_l(\mathbf{x})$ is the predicted value from the lth regression tree.

Boosting Boosting is also a general-purpose method for improving the accuracy of a predictive model. In boosting, trees are created sequentially by fitting a small tree each time to the residuals from the previous tree. The new tree is added to the previous tree with some weight and the residuals are recalculated. The process is repeated many times to obtain a final boosted model.

Random forests The random forest method is similar to the bagging method in that both methods build many trees and average these trees to create a final model. In the bagging method, all explanatory variables are used to fit trees. As a result, trees created in the bagging method might be highly correlated. In the random forest method, however, only a small subset of the explanatory variables is used to fit trees. For this reason, trees created in the random forest method might not be correlated, leading to less variance in the aggregated tree.

12.2 Implementation

To test the performance of regression trees for predicting fair market values, we will use the following R packages: `rpart`, `gbm`, and `randomForest`. The first package contains functions for fitting regression trees. The second package is used to create boosted tree models. The last package provides functions for creating bagged trees and random forests. If these packages have not been installed, you can use the function `install.packages` to install them. To install the package `rpart`, for example, we run the following code:

```
1 isntall.packages("rpart")
```

and then follow the instructions.

12.3   Applications

In this section, we test regression trees with the synthetic dataset described
in Appendix A. For our tests, we load and prepare the data as follows:

```
 1 > require(rpart)
 2 > require(gbm)
 3 > require(randomForest)
 4 > inforce <- read.csv("inforce.csv")
 5 >
 6 > vNames <- c("gbAmt", "gmwbBalance", "withdrawal",
       paste("FundValue", 1:10, sep=""))
 7 >
 8 > age <- with(inforce, (currentDate-birthDate)/365)
 9 > ttm <- with(inforce, (matDate - currentDate)/365)
10 >
11 > datN <- cbind(inforce[,vNames], data.frame(age=age,
       ttm=ttm))
12 > datC <- inforce[,c("gender", "productType")]
13 >
14 > greek <- read.csv("Greek.csv")
15 > greek <- greek[order(greek$recordID),]
16 >
17 > dat <- cbind(datN, datC, fmv=greek$fmv/1000)
18 > summary(dat)
19     gbAmt              gmwbBalance            withdrawal
20  Min.    : 50002   Min.    :      0   Min.    :      0
21  1st Qu.:179759   1st Qu.:      0   1st Qu.:      0
22  Median :303525   Median :      0   Median :      0
23  Mean    :313507   Mean    : 36141   Mean    : 21928
24  3rd Qu.:427544   3rd Qu.:      0   3rd Qu.:      0
25  Max.    :989205   Max.    :499709   Max.    :499586
26
27   FundValue1          FundValue2            FundValue3
28  Min.    :      0   Min.    :      0   Min.    :      0
29  1st Qu.:      0   1st Qu.:      0   1st Qu.:      0
30  Median :   8299   Median :   8394   Median :   4942
31  Mean    : 26611   Mean    : 26045   Mean    : 17391
32  3rd Qu.: 39209   3rd Qu.: 38463   3rd Qu.: 24251
33  Max.    :921549   Max.    :844323   Max.    :580753
34
35   FundValue4          FundValue5            FundValue6
36  Min.    :      0   Min.    :      0   Min.    :      0
37  1st Qu.:      0   1st Qu.:      0   1st Qu.:      0
38  Median :   4225   Median :   7248   Median :   8556
39  Mean    : 14507   Mean    : 21041   Mean    : 26570
40  3rd Qu.: 20756   3rd Qu.: 32112   3rd Qu.: 39241
41  Max.    :483937   Max.    :494382   Max.    :872707
```

```
42
43     FundValue7            FundValue8            FundValue9
44   Min.   :       0    Min.   :       0    Min.   :       0
45   1st Qu.:       0    1st Qu.:       0    1st Qu.:       0
46   Median :    6602    Median :    6255    Median :    5943
47   Mean   :   21506    Mean   :   19990    Mean   :   19647
48   3rd Qu.:   31088    3rd Qu.:   29404    3rd Qu.:   28100
49   Max.   :  634819    Max.   :  562485    Max.   :  663196
50
51     FundValue10              age                  ttm
52   Min.   :       0    Min.   :34.52    Min.   : 0.5863
53   1st Qu.:       0    1st Qu.:42.03    1st Qu.:10.3425
54   Median :    6738    Median :49.45    Median :14.5123
55   Mean   :   21003    Mean   :49.49    Mean   :14.5362
56   3rd Qu.:   31256    3rd Qu.:56.96    3rd Qu.:18.7616
57   Max.   :  599675    Max.   :64.46    Max.   :28.5205
58
59   gender          productType            fmv
60   F: 76007     ABRP    : 10000    Min.   : -69.938
61   M:113993     ABRU    : 10000    1st Qu.:   4.542
62                ABSU    : 10000    Median :  40.544
63                DBAB    : 10000    Mean   :  97.748
64                DBIB    : 10000    3rd Qu.: 108.141
65                DBMB    : 10000    Max.   :1784.549
66                (Other):130000
```

At the beginning, we load the three packages if they have not been loaded. We also combine the explanatory variables and the response variable (i.e., fmv) into one data frame.

12.3.1 Regression Trees with Latin Hypercube Sampling

In this section, we use representative policies selected by Latin hypercube sampling to train the model. To test the performance of a single regression tree, we proceed as follows:

```
 1 > S <- read.table("lhs.csv", sep=",")
 2 > S <- S[,2]
 3 >
 4 > {t1 <- proc.time()
 5 + set.seed(1)
 6 + tree1 <- rpart(fmv ~ ., data=dat[S,], method = "anova
       ")
 7 + yhat <- predict(tree1, dat)
 8 + proc.time()-t1}
 9    user  system elapsed
10    0.19    0.09    0.29
11 >
```

```
12 > calMeasure(greek$fmv/1000, yhat)
13 $pe
14 [1] -0.008948673
15
16 $r2
17 [1] 0.6503093
```

The validation measures show that the regression tree fits the data well. The percent error is within 1%. Figure 12.1 shows the scatter plot and the QQ plot produced by this model. From the figure, we see that there are only five distinct values among the predicted values. This means that the final regression tree has only five terminal nodes.

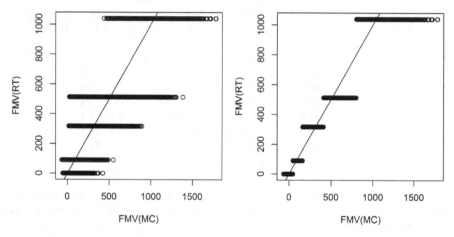

FIGURE 12.1: A scatter plot and a QQ plot of the fair market values calculated by Monte Carlo simulation and those predicted by a regression tree. The representative policies are selected by Latin hypercube sampling.

In tree-based models, we can get the variable importance based on which variable is used in each split. We can plot the variable importance in a bar plot by running the following code:

```
1 tree1summary <- summary(tree1)
2 vImp <- tree1summary$variable.importance
3 vImp <- vImp * 100 / max(vImp)
4 ind <- order(vImp)
5 dev.new(width=6, height=4)
6 par(las=2) # make label text perpendicular to axis
7 par(mar=c(3,6,1,1)) # increase y-axis margin.
8 barplot(vImp[ind], main="", beside=TRUE,horiz=TRUE,
        names.arg=names(vImp[ind]), cex.names=0.8)
```

After running the above code, we see the bar plot given in Figure 12.2. From the figure, we see that the most important variables are productType, gbAmt, and ttm. This is reasonable because productType determines how the guarantee payoffs are calculated and gbAmt determines the amount of guarantee payments.

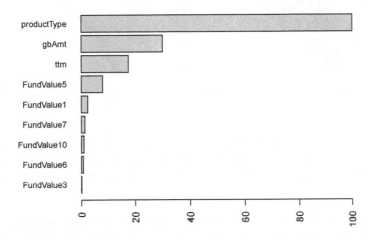

FIGURE 12.2: Variable importance obtained from the regression tree.

To create a bagged regression tree, we proceed as follows:

```
1 > {t1 <- proc.time()
2 + set.seed(1)
3 + bag1 <- randomForest(formula=fmv ~ ., data=dat[S,],
       mtry=17, importance=TRUE)
4 + yhat <- predict(bag1, dat)
5 + proc.time()-t1}
6     user   system elapsed
7     6.13     0.14    6.38
8 > calMeasure(greek$fmv/1000, yhat)
9 $pe
10 [1] -0.04182695
11
12 $r2
13 [1] 0.798825
```

In the above output, we see that the function randomForest was used to create a bagged tree. The reason is that the bagging method is a special case of the random forest method when all explanatory variables are used. We set mtry to 17 as we wanted to use all 17 explanatory variables. The R^2 is higher than that from the previous case.

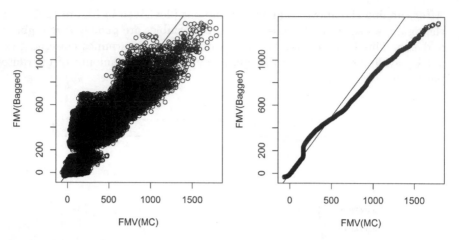

FIGURE 12.3: A scatter plot and a QQ plot of the fair market values calcu-lated by Monte Carlo simulation and those predicted by a bagged regression tree. The representative policies are selected by Latin hypercube sampling.

Figure 12.3 shows the scatter plot and the QQ plot produced from the bagged regression tree. Since a bagged tree is the average of many trees, the scatter plot and the QQ plot look much better than those based on a single tree. The QQ plot shows that the bagged model does not fit the right tail well.

The following output shows the performance of a boosted regression tree:

```
1 > {t1 <- proc.time()
2 + set.seed(1)
3 + boost1 <- gbm(formula=fmv ~ ., data=dat[S,],
      distribution="gaussian", n.trees=1000, interaction.
      depth=3)
4 + yhat <- predict(boost1, newdata=dat, n.trees=1000)
5 + proc.time()-t1}
6    user  system elapsed
7    5.62    0.02    5.67
8 > calMeasure(greek$fmv/1000, yhat)
9 $pe
10 [1]  -0.08728489
11
12 $r2
13 [1]  0.8173961
```

The R^2 is higher than that based on the bagged tree. However, the percentage error is larger in magnitude. The scatter plot and the QQ plot from the boosted tree are shown in Figure 12.4. The patterns are similar as those based on the bagged tree.

The following output shows the performance of a random forest:

```
1 > {t1 <- proc.time()
2 + set.seed(1)
```

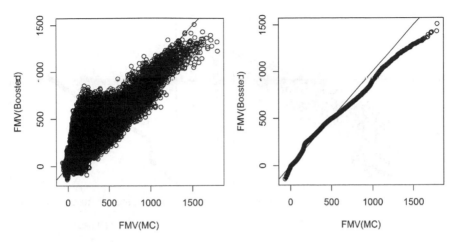

FIGURE 12.4: A scatter plot and a QQ plot of the fair market values calculated by Monte Carlo simulation and those predicted by a boosted regression tree. The representative policies are selected by Latin hypercube sampling.

```
 3 + bag1 <- randomForest(formula=fmv ~ ., data=dat[S,],
        mtry=4, importance=TRUE)
 4 + yhat <- predict(bag1, dat)
 5 + proc.time()-t1}
 6    user  system elapsed
 7    5.73    0.06    5.91
 8 > calMeasure(greek$fmv/1000, yhat)
 9 $pe
10 [1] -0.00664855
11
12 $r2
13 [1] 0.7430954
```

From the output, we see that the percentage error decreased. However, the R^2 also decreased. The scatter and QQ plots are shown in Figure 12.5. From the figure, we see that the random forest model did not fit the data well.

12.3.2 Regression Trees with Conditional Latin Hypercube Sampling

In this section, we test tree-based models when conditional Latin hypercube sampling is used to select representative policies.

First, we change the indices of the representative policies and test the regression tree model as follows:

```
 1 > S <- read.table("clhs.csv", sep=",")
 2 > S <- S[,2]
```

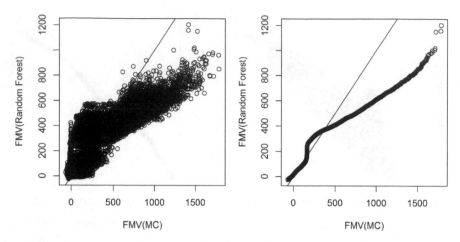

FIGURE 12.5: A scatter plot and a QQ plot of the fair market values calculated by Monte Carlo simulation and those predicted by a random forest. The representative policies are selected by Latin hypercube sampling.

```
 3  >
 4  > {t1 <- proc.time()
 5  + set.seed(1)
 6  + tree1 <- rpart(fmv ~ ., data=dat[S,], method = "anova
          ")
 7  + yhat <- predict(tree1, dat)
 8  + proc.time()-t1}
 9     user   system elapsed
10     0.19     0.06    0.25
11  >
12  > calMeasure(greek$fmv/1000, yhat)
13  $pe
14  [1] -0.06940123
15
16  $r2
17  [1] 0.7893949
```

The R^2 shows that the results are better than the case based on Latin hypercube sampling. In terms of percentage error, however, the results are worse than the case based on Latin hypercube sampling. Figure 12.6 shows the scatter plot and the QQ plot of this case. From the figure, we see that the regression tree has only six terminal nodes.

The following output shows the performance of a bagged tree:

```
 1  > {t1 <- proc.time()
 2  + set.seed(1)
 3  + bag1 <- randomForest(formula=fmv ~ ., data=dat[S,],
          mtry=17, importance=TRUE)
```

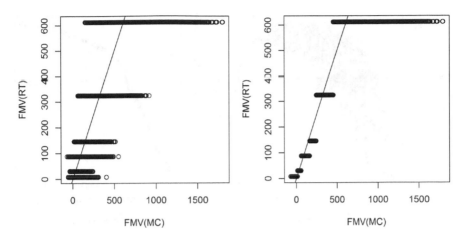

FIGURE 12.6: A scatter plot and a QQ plot of the fair market values calculated by Monte Carlo simulation and those predicted by a regression tree. The representative policies are selected by conditional Latin hypercube sampling.

```
4 + yhat <- predict(bag1, dat)
5 + proc.time()-t1}
6    user  system elapsed
7    6.47    0.16    6.65
8 > calMeasure(greek$fmv/1000, yhat)
9 $pe
10 [1] -0.03226062
11
12 $r2
13 [1] 0.8855023
```

The validation measures show that the results are better than the single tree model. The scatter plot and the QQ plot given in Figure 12.7 show that the model does not fit the right tail well.

The following output shows the performance of a boosted regression tree:

```
1 > {t1 <- proc.time()
2 + set.seed(1)
3 + boost1 <- gbm(formula=fmv ~ ., data=dat[S,],
        distribution="gaussian", n.trees=1000, interaction.
        depth=3)
4 + yhat <- predict(boost1, newdata=dat, n.trees=1000)
5 + proc.time()-t1}
6    user  system elapsed
7    5.47    0.02    5.69
8 > calMeasure(greek$fmv/1000, yhat)
9 $pe
10 [1] -0.008173688
```

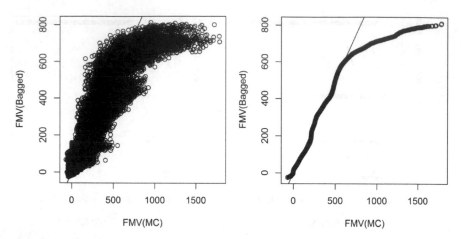

FIGURE 12.7: A scatter plot and a QQ plot of the fair market values calculated by Monte Carlo simulation and those predicted by a bagged regression tree. The representative policies are selected by conditional Latin hypercube sampling.

```
11
12 $r2
13 [1] 0.9024634
```

The validation measures show that the boosted model performed well. The scatter plot and the QQ plot of this case are given in Figure 12.7. We see similar patterns as in the previous case. The boosted model also does not fit the right tail well.

For random forests, we see the following results:

```
1 > {t1 <- proc.time()
2 + set.seed(1)
3 + bag1 <- randomForest(formula=fmv ~ ., data=dat[S,],
      mtry=4, importance=TRUE)
4 + yhat <- predict(bag1, dat)
5 + proc.time()-t1}
6    user   system elapsed
7    6.18     0.14    6.46
8 > calMeasure(greek$fmv/1000, yhat)
9 $pe
10 [1] -0.0150661
11
12 $r2
13 [1] 0.7444701
```

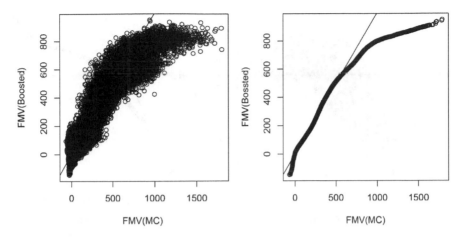

FIGURE 12.8: A scatter plot and a QQ plot of the fair market values calculated by Monte Carlo simulation and those predicted by a boosted regression tree. The representative policies are selected by conditional Latin hypercube sampling.

In terms of R^2, the random forest model is not as good as the bagged and the boosted models. Figure 12.9 shows the scatter plot and the QQ plot of this case. From the plots, we see that the predicted values of many policies are off.

12.3.3 Regression Trees with Hierarchical k-Means

In this section, we examine the performance of tree-based models when hierarchical k-means is used to select representative policies.

The following output shows the results of the single regression tree:

```
 1 > S <- read.table("hkmeans.csv", sep=",")
 2 > S <- S[,2]
 3 >
 4 > {t1 <- proc.time()
 5 + set.seed(1)
 6 + tree1 <- rpart(fmv ~ ., data=dat[S,],   method = "anova
      ")
 7 + yhat <- predict(tree1, dat)
 8 + proc.time()-t1}
 9    user   system elapsed
10    0.19    0.08    0.27
11 >
12 > calMeasure(greek$fmv/1000, yhat)
13 $pe
14 [1]  -0.0454955
15
```

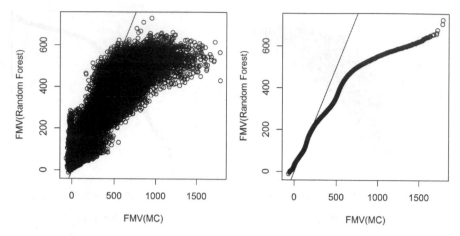

FIGURE 12.9: A scatter plot and a QQ plot of the fair market values calculated by Monte Carlo simulation and those predicted by a random forest. The representative policies are selected by conditional Latin hypercube sampling.

```
16 $r2
17 [1]  0.8107462
```

The R^2 is higher than the cases when Latin hypercube sampling and conditional Latin hypercube sampling were used to select representative policies. Figure 12.10 shows the scatter and the QQ plots of this case. We see similar patterns as before when a single tree was used to make predictions.

The following output shows the performance of a bagged tree:

```
1 > {t1 <- proc.time()
2 + set.seed(1)
3 + bag1 <- randomForest(formula=fmv ~ ., data=dat[S,],
      mtry=17, importance=TRUE)
4 + yhat <- predict(bag1, dat)
5 + proc.time()-t1}
6    user   system elapsed
7    6.41     0.33    6.82
8 > calMeasure(greek$fmv/1000, yhat)
9 $pe
10 [1]  0.04922809
11
12 $r2
13 [1]  0.8681998
```

From the output, we see that the R^2 improved from the previous case. The scatter plot and the QQ plot are shown in Figure 12.11, which shows that the model does not fit the right tail well.

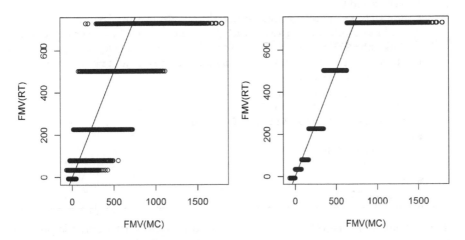

FIGURE 12.10: A scatter plot and a QQ plot of the fair market values calculated by Monte Carlo simulation and those predicted by a regression tree.
The representative policies are selected by hierarchical k-means.

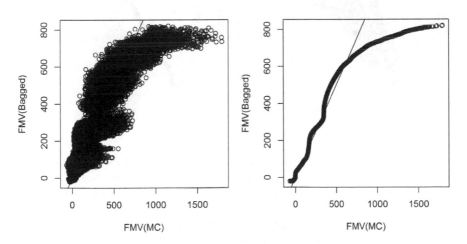

FIGURE 12.11: A scatter plot and a QQ plot of the fair market values calculated by Monte Carlo simulation and those predicted by a bagged regression
tree. The representative policies are selected by hierarchical k-means.

```
 1 > {t1 <- proc.time()
 2 + set.seed(1)
 3 + boost1 <- gbm(formula=fmv ~ ., data=dat[S,],
       distribution="gaussian", n.trees=1000, interaction.
       depth=3)
 4 + yhat <- predict(boost1, newdata=dat, n.trees=1000)
 5 + proc.time()-t1}
 6     user   system elapsed
 7     5.65     0.03    5.80
 8 > calMeasure(greek$fmv/1000, yhat)
 9 $pe
10 [1] 0.01030868
11
12 $r2
13 [1] 0.882043
```

The validation measures show that the results are better than those produced by the bagged tree in the previous case. Figure 12.12 shows the scatter and the QQ plots, which look better than those in the previous case.

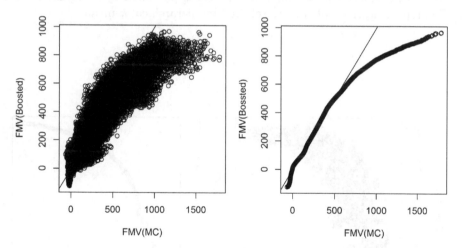

FIGURE 12.12: A scatter plot and a QQ plot of the fair market values calculated by Monte Carlo simulation and those predicted by a boosted regression tree. The representative policies are selected by hierarchical k-means.

The performance of a random forest is shown below:

```
 1 > {t1 <- proc.time()
 2 + set.seed(1)
 3 + bag1 <- randomForest(formula=fmv ~ ., data=dat[S,],
       mtry=4, importance=TRUE)
 4 + yhat <- predict(bag1, dat)
 5 + proc.time()-t1}
 6     user   system elapsed
```

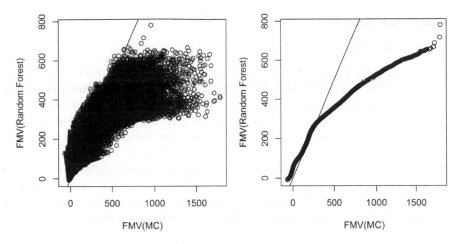

FIGURE 12.13: A scatter plot and a QQ plot of the fair market values calculated by Monte Carlo simulation and those predicted by a random forest. The representative policies are selected by hierarchical k-means.

```
 7      5.92      0.05      5.96
 8 > calMeasure(greek$fmv/1000, yhat)
 9 $pe
10 [1] 0.213374
11
12 $r2
13 [1] 0.6509478
```

The validation measures show that the random forest does not fit the data well. The percentage error is above 20%. Figure 12.13 shows the scatter and the QQ plots, which show that the random forest does not fit the data well.

TABLE 12.1: Validation measures of tree-based models with different experimental design methods.

	PE			R^2		
	LHS	cLHS	Hkmeans	LHS	cLHS	Hkmeans
Single	−0.89%	−6.94%	−4.55%	0.6503	0.7894	0.8107
Bagged	−4.18%	−3.23%	4.92%	0.7988	0.8855	0.8682
Boosted	−8.73%	−0.82%	1.03%	0.8174	0.9025	0.8820
Random Forest	-0.66%	−1.51%	21.34%	0.7431	0.7445	0.6509

12.4 Summary

In this chapter, we introduced tree-based models and illustrated their use in predicting fair market values for a large portfolio of variable annuity contracts. Tree-based models have some advantages over other models. For example, tree-based models can handle categorical variables straightforwardly. For more information about tree-based models, readers are referred to Breiman et al. (1984), Mitchell (1997), Gan et al. (2018), and Quan et al. (2018).

Table 12.1 summarized the performance of tree-based models with different experimental design methods. From the table, we see that boosted models work the best, especially when conditional Latin hypercube sampling and hierarchical k-means were used to select representative polices.

A

Synthetic Datasets

In this appendix, we describe synthetic datasets that can be used to study the computational problems associated with variable annuity valuation. These synthetic datasets were created by and explained in Gan and Valdez (2017b). In that paper, readers can learn the details of how the variable annuity policies are generated and how Monte Carlo simulation was used to value these policies.

A.1   The Synthetic Inforce

The synthetic dataset contains 190,000 variable annuity policies, each of which is described by 45 features that are shown in Table A.1. Most of the features are related to investment funds, which are described by 10 fund values, 10 fund numbers, and 10 fund fees. There are 19 different variable annuity products in the synthetic inforce. These products are described in Table A.2. Each product type has 10,000 policies.

About 40% policies in the synthetic dataset have female policyholders. Table A.3 shows the distribution of gender by product type. Table A.4 shows the summary statistics of the age, the time to maturity, and the dollar fields. The age is the duration in years between the birth date and the current date. The time to maturity is the duration in years between the current date and the maturity date.

A.2   The Greeks

Table A.5 shows some summary statistics of the fair market values and Greeks of individual policies that are calculated by Monte Carlo simulation. From Table A.5, we see that some policies have negative fair market values. For these policies, the guarantee benefit payoff is less than the risk charge. From the table, we also see that some contracts have positive deltas. The policies that have positive deltas are contracts with the annual ratchet guarantee benefit.

TABLE A.1: A list of variable annuity policy features.

Feature	Description
recordID	Unique identifier of the policy
survivorShip	Positive weighting number
gender	Gender of the policyholder
productType	Product type
issueDate	Issue date
matDate	Maturity date
birthDate	Birth date of the policyholder
currentDate	Current date
baseFee	M&E (Mortality & Expense) fee
riderFee	Rider fee
rollUpRate	Roll-up rate
gbAmt	Guaranteed benefit
gmwbBalance	GMWB balance
wbWithdrawalRate	Guaranteed withdrawal rate
withdrawal	Withdrawal so far
FundValue$i$	Fund value of the ith investment fund
FundNum$i$	Fund number of the ith investment fund
FundFee$i$	Fund management fee of the ith investment fund

For such contracts, the value of the guarantee may increase when the market goes up because the guarantee benefit is reset to the maximum of the current guarantee benefit and the account value if the latter is higher.

Figures A.1 and A.2 show the histograms of the fair market values and the partial Greeks of individual policies. From Figure A.1, we see that the distribution of the fair market values is highly skewed with a long right tail. The distributions of the partial deltas are skewed to the left. Figure A.2 shows histograms of partial dollar rhos. In particular, the distribution of the 30-year rho is skewed to the left and has a long left tail.

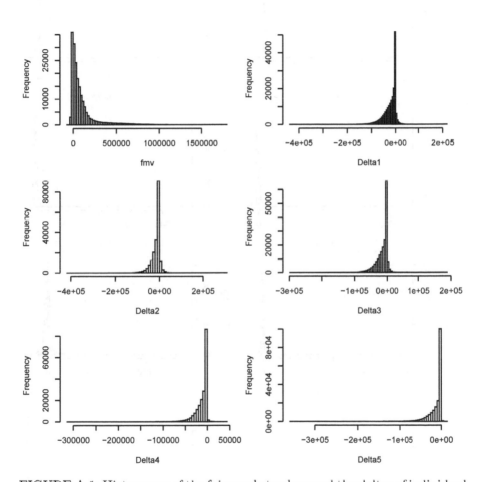

FIGURE A.1: Histograms of the fair market values and the deltas of individual policies.

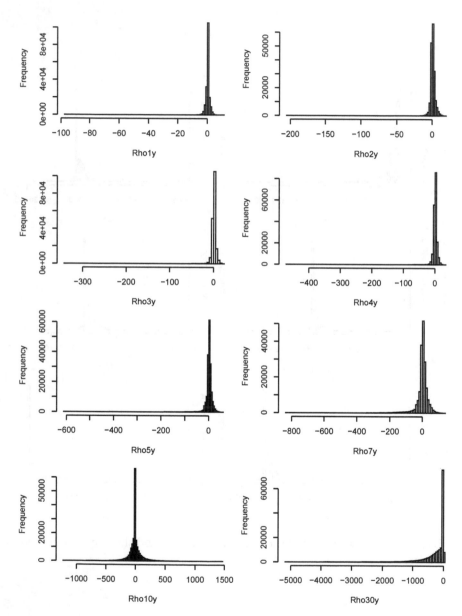

FIGURE A.2: Histograms of the rhos of individual policies.

TABLE A.2: Types of variable annuity contracts contained in the synthetic portfolio.

Product	Description
DBRP	GMDB with return of premium
DBRU	GMDB with annual roll-up
DBSU	GMDB with annual ratchet
ABRP	GMAB with return of premium
ABRU	GMAB with annual roll-up
ABSU	GMAB with annual ratchet
IBRP	GMIB with return of premium
IBRU	GMIB with annual roll-up
IBSU	GMIB with annual ratchet
MBRP	GMMB with return of premium
MBRU	GMMB with annual roll-up
MBSU	GMMB with annual ratchet
WBRP	GMWB with return of premium
WBRU	GMWB with annual roll-up
WBSU	GMWB with annual ratchet
DBAB	GMDB + GMAB with annual ratchet
DBIB	GMDB + GMIB with annual ratchet
DBMB	GMDB + GMMB with annual ratchet
DBWB	GMDB + GMWB with annual ratchet

TABLE A.3: Distribution of gender by product type.

productType	Female	Male	Total	productType	Female	Male	Total
ABRP	4068	5932	10000	IBRP	4007	5993	10000
ABRU	3974	6026	10000	IBRU	4027	5973	10000
ABSU	4054	5946	10000	IBSU	4007	5993	10000
DBAB	3974	6026	10000	MBRP	3909	6091	10000
DBIB	3948	6052	10000	MBRU	3992	6008	10000
DBMB	4013	5987	10000	MBSU	3980	6020	10000
DBRP	4002	5998	10000	WBRP	3970	6030	10000
DBRU	3952	6048	10000	WBRU	4076	5924	10000
DBSU	4038	5962	10000	WBSU	3994	6006	10000
DBWB	4022	5978	10000				

TABLE A.4: Summary statistics of continuous features. Features `age` and `ttm` are calculated from the birth date, valuation date, and maturity date.

	Min	1st Q	Mean	3rd Q	Max
gbAmt	50,001.72	179,758.97	313,507.22	427,544.13	989,204.53
gmwbBalance	0.00	0.00	36,140.74	0.00	499,708.73
withdrawal	0.00	0.00	21,927.80	0.00	499,585.73
FundValue1	0.00	0.00	26,611.38	39,208.90	921,548.70
FundValue2	0.00	0.00	26,044.48	38,463.42	844,322.70
FundValue3	0.00	0.00	17,391.42	24,251.44	580,753.42
FundValue4	0.00	0.00	14,507.35	20,755.88	483,936.90
FundValue5	0.00	0.00	21,041.04	32,111.72	494,381.61
FundValue6	0.00	0.00	26,569.62	39,241.12	872,706.64
FundValue7	0.00	0.00	21,505.81	31,087.78	634,819.08
FundValue8	0.00	0.00	19,990.40	29,404.16	562,485.37
FundValue9	0.00	0.00	19,646.92	28,100.22	663,196.22
FundValue10	0.00	0.00	21,002.82	31,255.73	599,675.34
age	34.52	42.03	49.49	56.96	64.46
ttm	0.59	10.34	14.54	18.76	28.52

TABLE A.5: Summary statistics of the fair market values and Greeks of individual policies.

	Min	1st Q	Mean	3rd Q	Max
FMV	−69,938.25	4,541.54	97,747.87	108,140.94	1,784,549.09
Delta1	−435,069.76	−34,584.04	−22,267.27	−2,218.74	216,124.52
Delta2	−412,495.69	−21,234.31	−13,698.78	−493.09	300,018.51
Delta3	−296,034.74	−22,704.23	−15,022.28	−865.73	187,414.30
Delta4	−312,777.09	−16,568.88	−11,598.56	−655.58	40,923.66
Delta5	−355,741.29	−16,517.41	−12,325.23	0.00	10,391.24
Rho1y	−97.76	−0.01	0.21	0.67	11.44
Rho2y	−204.80	0.00	0.88	2.29	19.63
Rho3y	−343.17	−0.03	0.45	2.70	23.23
Rho4y	−472.70	−0.06	0.02	3.42	31.82
Rho5y	−610.45	−0.38	−0.51	6.56	63.00
Rho7y	−834.00	−4.63	−2.87	14.35	136.84
Rho10y	−1,236.08	−40.65	−7.41	16.11	1,472.07
Rho30y	−5,158.97	−388.03	−327.03	0.00	0.00

Bibliography

Barton, R. R. (2015). Tutorial: Simulation metamodeling. In *Proceedings of the 2015 Winter Simulation Conference*, pages 1765–1779, Piscataway, NJ. IEEE Press.

Box, G. E. P. and Draper, N. R. (2007). *Response Surfaces, Mixtures, and Ridge Analyses*. Wiley, Hoboken, NJ, 2nd edition.

Breiman, L., Friedman, J., Stone, C. J., and Olshen, R. (1984). *Classification and Regression Trees*. Chapman & Hall/CRC, Raton Boca, FL.

Cathcart, M. J., Lok, H. Y., McNeil, A. J., and Morrison, S. (2015). Calculating variable annuity liability "greeks" using Monte Carlo simulation. *ASTIN Bulletin*, 45(2):239 – 266.

Chueh, Y. C. M. (2002). Efficient stochastic modeling for large and consolidated insurance business: Interest rate sampling algorithms. *North American Actuarial Journal*, 6(3):88–103.

Cressie, N. (1993). *Statistics for Spatial Data*. Wiley, Hoboken, NJ, revised edition.

Cummins, J., Dionne, G., McDonald, J. B., and Pritchett, B. (1990). Applications of the GB2 family of distributions in modeling insurance loss processes. *Insurance: Mathematics and Economics*, 9(4):257 – 272.

Dardis, T. (2016). Model efficiency in the U.S. life insurance industry. *The Modeling Platform*, (3):9–16.

Das, R. N. (2014). *Robust Response Surfaces, Regression, and Positive Data Analyses*. CRC Press, Boca Raton, FL.

Doyle, D. and Groendyke, C. (2018). Using neural networks to price and hedge variable annuity guarantees. *Risks*, 7(1).

Feng, R. (2018). *An Introduction to Computational Risk Management of Equity-Linked Insurance*. CRC Press, Boca Raton, FL.

Friedman, L. W. (1996). *The Simulation Metamodel*. Kluwer Academic Publishers, Norwell, MA, USA.

Gan, G. (2011). *Data Clustering in C++: An Object-Oriented Approach.* Data Mining and Knowledge Discovery Series. Chapman & Hall/CRC Press, Boca Raton, FL, USA.

Gan, G. (2013). Application of data clustering and machine learning in variable annuity valuation. *Insurance: Mathematics and Economics*, 53(3):795–801.

Gan, G. (2015). Application of metamodeling to the valuation of large variable annuity portfolios. In *Proceedings of the Winter Simulation Conference*, pages 1103–1114.

Gan, G. (2018). Valuation of large variable annuity portfolios using linear models with interactions. *Risks*, 6(3):71.

Gan, G. and Huang, J. (2017). A data mining framework for valuing large portfolios of variable annuities. In *Proceedings of the 23rd ACM SIGKDD International Conference on Knowledge Discovery and Data Mining*, pages 1467–1475.

Gan, G., Lan, Q., and Ma, C. (2016). Scalable clustering by truncated fuzzy c-means. *Big Data and Information Analytics*, 1(2/3):247–259.

Gan, G. and Lin, X. S. (2015). Valuation of large variable annuity portfolios under nested simulation: A functional data approach. *Insurance: Mathematics and Economics*, 62:138 – 150.

Gan, G. and Lin, X. S. (2017). Efficient greek calculation of variable annuity portfolios for dynamic hedging: A two-level metamodeling approach. *North American Actuarial Journal*, 21(2):161–177.

Gan, G., Quan, Z., and Valdez, E. A. (2018). Machine learning and its application in variable annuity valuation. In *Proceedings of the 2018 SIAM International Conference on Data Mining*.

Gan, G. and Valdez, E. A. (2016). An empirical comparison of some experimental designs for the valuation of large variable annuity portfolios. *Dependence Modeling*, 4(1):382–400.

Gan, G. and Valdez, E. A. (2017a). Modeling partial greeks of variable annuities with dependence. *Insurance: Mathematics and Economics*, 76:118–134.

Gan, G. and Valdez, E. A. (2017b). Valuation of large variable annuity portfolios: Monte Carlo simulation and synthetic datasets. *Dependence Modeling*, 5:354–374.

Gan, G. and Valdez, E. A. (2018a). Data clustering with actuarial applications. *North American Actuarial Journal*. Accepted.

Gan, G. and Valdez, E. A. (2018b). Nested stochastic valuation of large variable annuity portfolios: Monte Carlo simulation and synthetic datasets. *Data*, 3(3):31.

Gan, G. and Valdez, E. A. (2018c). Regression modeling for the valuation of large variable annuity portfolios. *North American Actuarial Journal*, 22(1):40–54.

Gan, G. and Valdez, E. A. (2018d). Valuation of large variable annuity portfolios with rank order kriging. Submitted for publication.

Hardy, M. (2003). *Investment Guarantees: Modeling and Risk Management for Equity-Linked Life Insurance*. John Wiley & Sons, Inc., Hoboken, New Jersey.

Hejazi, S. A. and Jackson, K. R. (2016). A neural network approach to efficient valuation of large portfolios of variable annuities. *Insurance: Mathematics and Economics*, 70:169 – 181.

Hejazi, S. A., Jackson, K. R., and Gan, G. (2017). A spatial interpolation framework for efficient valuation of large portfolios of variable annuities. *Quantitative Finance and Economics*, 1(2):125–144.

Huang, Z. (1998). Extensions to the k-means algorithm for clustering large data sets with categorical values. *Data Mining and Knowledge Discovery*, 2(3):283–304.

Isaaks, E. and Srivastava, R. (1990). *An Introduction to Applied Geostatistics*. Oxford University Press, Oxford, UK.

Kleijnen, J. P. (2009). Kriging metamodeling in simulation: A review. *European Journal of Operational Research*, 192(3):707 – 716.

Kleijnen, J. P. C. (1975). A comment on Blanning's "metamodel for sensitivity analysis: The regression metamodel in simulation". *Interfaces*, 5(3):21–23.

Klugman, S., Panjer, H., and Willmot, G. (2012). *Loss Models: From Data to Decisions*. Wiley, Hoboken, NJ, 4th edition.

Ledlie, M. C., Corry, D. P., Finkelstein, G. S., Ritchie, A. J., Su, K., and Wilson, D. C. E. (2008). Variable annuities. *British Actuarial Journal*, 14(2):327–389.

Lim, M. and Hastie, T. J. (2015). Learning interactions via hierarchical group-lasso regularization. *Journal of Computational and Graphical Statistics*, 24(3):627–654.

Loeppky, J. L., Sacks, J., and Welch, W. J. (2009). Choosing the sample size of a computer experiment: A practical guide. *Technometrics*, 51(4):366–376.

Longley-Cook, A. G. (2003). Efficient stochastic modeling utilizing representative scenarios: Application to equity risks. In *Stochastic Modeling Symposium*.

Maclean, J. (1962). *Life Insurance*. McGraw Hill, New York, NY, 9th edition.

MacQueen, J. (1967). Some methods for classification and analysis of multivariate observations. In LeCam, L. and Neyman, J., editors, *Proceedings of the 5th Berkeley Symposium on Mathematical Statistics and Probability*, volume 1, pages 281–297, Berkeley, CA, USA. University of California Press.

McKay, B. and Wanless, I. (2008). A census of small Latin hypercubes. *SIAM Journal on Discrete Mathematics*, 22(2):719–736.

McKay, M., Beckman, R., and Conover, W. J. (1979). A comparison of three methods for selecting values of input variables in the analysis of output from a computer code. *Technometrics*, 21(2):239–245.

Minasny, B. and McBratney, A. B. (2006). A conditioned Latin hypercube method for sampling in the presence of ancillary information. *Computers & Geosciences*, 32(9):1378 – 1388.

Mitchell, T. M. (1997). *Machine Learning*. McGraw-Hill.

Nister, D. and Stewenius, H. (2006). Scalable recognition with a vocabulary tree. In *2006 IEEE Computer Society Conference on Computer Vision and Pattern Recognition (CVPR'06)*, volume 2, pages 2161–2168.

O'Hagan, A. and Ferrari, C. (2017). Model-based and nonparametric approaches to clustering for data compression in actuarial applications. *North American Actuarial Journal*, 21(1):107–146.

Petelet, M., Iooss, B., Asserin, O., and Loredo, A. (2010). Latin hypercube sampling with inequality constraints. *AStA Advances in Statistical Analysis*, 94(4):325–339.

Phillips, P. (2012). Lessons learned about leveraging high performance computing for variable annuities. In *Equity-Based Insurance Guarantees Conference*, Chicago, IL.

Pistone, G. and Vicario, G. (2010). Comparing and generating Latin hypercube designs in kriging models. *AStA Advances in Statistical Analysis*, 94(4):353–366.

Poterba, J. (1997). The history of annuities in the United States. National Bureau of Economic Research (NBER) Working Paper Series, Working Paper 6001.

Quan, Z., Gan, G., and Valdez, E. A. (2018). Tree-based models for variable annuity valuation: Parameter tuning and empirical analysis. Submitted for publication.

Reynolds, C. and Man, S. (2008). Nested stochastic pricing: The time has come. *Product Matters! - Society of Actuaries*, 71:16–20.

Rosner, B. B. (2011). Model efficiency study results. Society of Actuaries.

Roudier, P. (2011). *clhs: a R package for conditioned Latin hypercube sampling*.

Sarukkali, M. (2013). *Replicated Stratified Sampling for Sensitivity Analysis*. PhD thesis, Department of Mathematics, University of Connecticut.

Sebestyen, G. S. (1962). Pattern recognition by an adaptive process of sample set construction. *IRE Transactions on Information Theory*, 8(5):82–91.

Sun, J., Frees, E. W., and Rosenberg, M. A. (2008). Heavy-tailed longitudinal data modeling using copulas. *Insurance: Mathematics and Economics*, 42(2):817 – 830.

The Geneva Association (2013). Variable annuities - an analysis of financial stability. Available online at: https://www.genevaassociation.org/media/618236/ga2013-variable_annuities.pdf.

Tibshirani, R. (1996). Regression shrinkage and selection via the lasso. *Journal of the Royal Statistical Society. Series B (Methodological)*, 58(1):267–288.

Vadiveloo, J. (2012). Replicated stratified sampling - A new financial modeling option. *Actuarial Research Clearing House*, 1:1–4.

Viana, F. (2013). Things you wanted to know about the Latin hypercube design and were afraid to ask. In *10th World Congress on Structural and Multidisciplinary Optimization*, Orlando, FL.

Wackernagel, H. (2003). *Multivariate Geostatistics: An Introduction with Applications*. Springer-Verlag, New York, NY, 3rd edition.

Xu, W., Chen, Y., Coleman, C., and Coleman, T. F. (2018). Moment matching machine learning methods for risk management of large variable annuity portfolios. *Journal of Economic Dynamics and Control*, 87:1 – 20.

Yamamoto, J. K. (2005). Correcting the smoothing effect of ordinary kriging estimates. *Mathematical Geology*, 37(1):69–94.

Yuan, M. and Lin, Y. (2006). Model selection and estimation in regression with grouped variables. *Journal of the Royal Statistical Society, Series B*, 68:49–67.

Index

Milton Keynes UK
Ingram Content Group UK Ltd.
UKHW040058071024
449327UK00019B/634